D0262520

50 ideas

you really need to know

the human brain

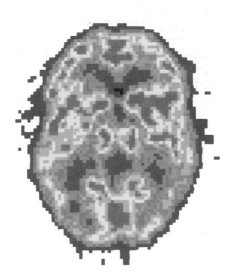

Moheb Costandi

Quercus

Contents

Introduction

Modern neuroscience can be traced back to the 1890s, when researchers first determined that the nervous system, like all other living things, is made of cells. Fast-forward a century: President George W. Bush declared the 1990s the 'Decade of the Brain', and since then research into the workings of this extremely complex organ has accelerated at an astonishing pace. Some say that we have learned more about the brain in the past decade than we did in the hundred years preceding it. Even so, we are only just beginning to scratch the surface, and a huge amount remains to be discovered.

During this short time, many theories about how the brain works, and how it generates our thought and behaviours, have been put forward. Some of the earlier ideas, such as phrenology – the 19th-century discipline that tried to correlate personality traits with the shape of the head – became influential in their time but were eventually debunked as pseudo-science. Others, such as the neuron doctrine – the idea that the brain is made of cells – remain central to modern neuroscience.

As technology advances and our understanding of the brain improves, the general public has become increasingly interested in neuroscience and in what these exciting new findings mean for them. At the same time, there is a great deal of sensationalism surrounding brain research, not to mention much inaccurate reporting.

Likewise, myths about the brain abound, and some of the more popular examples – such as the idea that the left brain is 'logical' and the right brain 'creative' – seem to be gaining traction, especially within education and the business sector.

This book is an attempt to distil over 100 years of thinking about the brain. It draws together influential ideas in neuroscience, updating old concepts in the light of new evidence, as well as introducing others that have emerged only recently. It attempts to explain these ideas accurately, and in a way that is easily digestible; to separate the wheat from the chaff; and to demystify the mysterious matter inside our heads. Where possible, I discuss how the science is carried out – the techniques used, and how researchers refine their ideas as new evidence emerges.

Some believe that gaining a better understanding of how the brain works will provide answers to life's big questions. It will not: brain research cannot tell us everything about ourselves, or what it means to be human. But it does offer the possibility that treatments may be developed for numerous debilitating conditions that afflict us, such as addiction, Alzheimer's disease, stroke and paralysis. *50 Human Brain Ideas You Really Need to Know* addresses these hopes, too, with cautious optimism.

01 The nervous system

The nervous system consists of two main components. One part, the central nervous system – made up of the brain and spinal cord – receives information from, and integrates the activity of, the rest of the body. The other section, the peripheral nervous system, contains nerves that send and receive information to and from the body.

The human brain contains hundreds of billions of cells arranged in a highly organized fashion, and is often said to be the most complex structure in the known universe – yet it weighs about only 1.5kg (just over 3lb). It consists of two hemispheres, each of which controls, and receives information from, the opposite side of the body. The cerebral cortex covering each hemisphere is divided into four specialized lobes; these all perform different functions, and are separated from one another by deep grooves called fissures.

The frontal lobe performs complex mental functions such as reasoning and decision-making, and also contains the motor areas, which plan and execute voluntary movements.

The parietal lobe contains the somatosensory areas, the parts that process touch information from the body. It also integrates different types of sensory information for spatial awareness – basically, knowledge of how the body is oriented within space.

TIMELINE

1700 BC	900	1543
The Edwin Smith papyrus, containing the first description of the nervous system, is written	Al-Razi describes the cranial nerves in *Kitab al-Hawi Fi Al Tibb*	Publication of *On the Workings of the Human Body* by Andreas Vesalius

Layers of complexity

The cortex (or 'bark') is the highly convoluted sheet of folded tissue that sits prominently on the outside of the brain. The cortex of human beings covers a much larger surface area than in other animals – extending 0.2m^2 (2½ sq ft) if laid out flat. This folding gives the cortex its familiar appearance, with its numerous gyri (bulges) and sulci (furrows). The cortex is just a few millimetres thick, but comprises six layers, with cells arranged uniformly in each layer. Despite this uniform structure, the cortex contains a large number of discrete areas, each specialized to perform a particular function.

The temporal lobe receives information from the ears, and its outer surface contains areas specialized for understanding speech. The inner surface contains the hippocampus, which is critical for memory formation and, with the surrounding areas, plays an important role in spatial navigation.

The occipital lobe is located at the back of the brain and contains dozens of distinct regions specialized to process and interpret visual information.

THE BRAIN UNCOVERED

Beneath the cortex lie several large clusters of neurons. The thalamus (or 'deep chamber') lies right at the brain's centre, and relays information from the sense organs to the appropriate region of the brain. Surrounding the thalamus are the basal ganglia, a group of structures involved mainly in the control of voluntary movement. The limbic system is another set of subcortical structures located between the basal ganglia and cortex. Sometimes referred to as the 'reptilian brain', the limbic system is evolutionarily primitive, and is involved in emotion, reward and motivation. It also includes the hippocampus and amygdala, both of which play a part in memory.

1641
Franciscus de la Boë Sylvius describes the fissure on the side of the brain

1664
Thomas Willis publishes *Cerebri anatome*

1695
Publication of Humphrey Ridley's *The Anatomy of the Brain*

The midbrain is a small area lying at the top of the brain stem. It contains clusters of neurons that control eye movement and is the main source of the neurotransmitter dopamine. Neurons that make dopamine also produce a pigment called melatonin, giving part of the midbrain a black appearance. This part of the midbrain is therefore called the substantia nigra ('black substance').

> ❝THE HUMAN BRAIN ... IS THE MOST COMPLICATED ORGANIZATION OF MATTER THAT WE KNOW.❞
>
> Isaac Asimov, 1986

The hindbrain comprises three structures located at the top of the spinal cord, which together make up the brain stem. The lower portion of the brain stem, the medulla oblongata, controls vital involuntary functions such as breathing and heart rate, and is closely associated with arousal. Above the medulla is the pons (or 'bridge'), connecting the cerebral cortex and the spinal cord and also linked to arousal. The third component of the hindbrain, the cerebellum (or 'little brain'), is involved in controlling balance and coordinating movement. It is essential for learning motor skills such as riding a bike, but is also associated with emotions and thought processes.

RUSH-HOUR TRAFFIC

The spinal cord, at the heart of the body's essential transport network, is an enormous bundle of millions of nerve fibres relaying information back and forth between the brain and the body. This very fragile structure, protected by the vertebral column, can perform certain functions, such as the knee-jerk reflex, on its own without any commands from the brain. It is segmented, with nerves leaving and entering it at regular intervals and in a highly orderly manner (in cross-section, it resembles a butterfly).

Motor neuron fibres leave the front of the spinal cord and extend to the body muscles, sending them information from the brain about voluntary movement. The axons of the sensory neurons carry information from the body into the back of the spinal cord, and form connections with second-order neurons that relay the information up into the brain. The axons of motor and sensory neurons are bundled together in the peripheral nerves.

MESSAGE CARRIERS

The peripheral nervous system consists of all the nerves that emanate from the brain and spinal cord, and consists of two different components. One, the somatic nervous system, is composed of the sensory and motor nerve fibres that carry information between the body and spinal cord. These nerves are involved with bodily sensations and the control of voluntary movement.

The other component is the autonomic nervous system, which controls the heart, glands and the smooth muscles in the blood vessels, eyes and gut, which are not under voluntary control.

> **MIND, A MYSTERIOUS FORM OF MATTER SECRETED BY THE BRAIN.**
> Ambrose Bierce, 1911

The autonomic nervous system can be further subdivided into the sympathetic and parasympathetic nervous systems, which have opposing functions. The sympathetic nervous system uses the neurotransmitter noradrenaline to increase heart rate, dilate the pupils and breathing tubes, and divert blood away from the digestive system. These effects prepare the body for action, in what is called the 'fight-or-flight' response. The para-sympathetic nervous system, on the other hand, uses the neurotransmitter acetylcholine to constrict the pupils and breathing tubes, slow the heart rate, and increase digestive function.

The cranial nerves are also part of the peripheral nervous system. These nerves emanate from the brain stem, and relay information between the brain and the sense organs. The vagus nerve, or tenth cranial nerve, is the longest of all, and sends branches as far as the heart, chest and abdomen.

The condensed idea
The nervous system is extremely complex and highly ordered

02 The neuron doctrine

Modern neuroscience is largely based on the idea that the brain is made of cells. The human brain is estimated to contain a staggering 80 to 120 billion neurons, which form intricate networks that process information. Neurons (nerve cells), one of the two types of brain cells, are specialized to produce electrical signals and communicate with each other.

I n the 1830s, two German scientists proposed cell theory, which states that all living things are made of cells. At that time, microscopes were not powerful enough to show the structure of the nervous system in any great detail, so it was unclear whether or not cell theory applied to nervous tissue, and this became the subject of long-lasting debate. Some researchers believed the nervous system must, like other parts of the body, also consist of cells, but others argued that it was composed of a continuous network of tissue.

As microscopes became more powerful and chemical staining methods improved, researchers began to see nervous tissue in increasingly finer detail. One important advance, made by Camillo Golgi, was the discovery of the so-called 'black reaction', a staining technique that involves hardening the tissue with potassium bichromate and ammonia, then immersing it in silver nitrate. The black reaction randomly stains small numbers of neurons in a tissue sample; the cells are stained in their entirety, making their shape visible in silhouette. In the 1880s, the Spanish neuroanatomist Santiago Ramón y Cajal

TIMELINE

1655	1838	1839	1865
Robert Hooke discovers cells	Robert Remak suggests that nerve fibres are joined to nerve cells	Theodor Schwann and Matthias Schleiden propose cell theory	Posthumous publication of Otto Deiters' description of axons and dendrites

started using Golgi's staining method to examine and compare tissue from many brain regions of various animal species. Cajal improved on the method by immersing his samples twice in the solutions. This stained the neurons more deeply, enabling him to study them in even greater detail.

> **LIKE THE ENTOMOLOGIST HUNTING FOR BRIGHTLY COLOURED BUTTERFLIES, MY ATTENTION WAS DRAWN TO THE FLOWER GARDEN OF THE GREY MATTER, WHICH CONTAINED CELLS WITH DELICATE AND ELEGANT FORMS, THE MYSTERIOUS BUTTERFLIES OF THE SOUL.**
>
> Santiago Ramón y Cajal, 1894

Cajal concluded that the brain is indeed made up of cells and, after convincing others that this was the case during a conference in 1889, the neuron doctrine – which states that neurons are the fundamental structural and functional units of the nervous system – was born, with Cajal and Golgi sharing the 1906 Nobel Prize in Physiology for their contributions. Despite inventing the method that led to the discovery of the neuron, Golgi somewhat ironically held on to the idea that the nervous system is composed of a continuous network of tissue. Cajal, on the other hand, is widely regarded as the father of modern neuroscience.

THE BODY'S MESSENGERS

The human brain contains at least several hundred, and perhaps as many as several thousand, different types of neurons, which come in many different shapes and sizes but which can be broadly classified into three types according to their function. Sensory neurons carry information from the sense organs into the brain; motor neurons send commands to the muscles and organs; and interneurons relay information between neurons in localized circuits, or over greater distances between neurons in different regions of the brain.

Despite this bewildering diversity, the vast majority of neurons share a number of basic features. Traditionally, neurons are subdivided into three 'compartments', each of which has a distinct function:

1873	**1889**	**2005**
Camillo Golgi discovers the 'black reaction'	Santiago Cajal argues that the nervous system consists of cells	Itzhak Fried and colleagues discover 'Jennifer Aniston' neurons

Jennifer Aniston cells

Researchers have discovered neurons that respond very specifically to images of well-known celebrities such as Jennifer Aniston or Halle Berry, or to famous landmarks such as the Eiffel Tower or the White House, while examining the brains of epileptic patients about to undergo neurosurgery. These cells are located in a part of the brain containing structures known to be critical for memory. Subsequently, the same researchers discovered that these cells are activated not only when the patients view images of the celebrities or landmarks, but also when they are merely thinking about them. These discoveries led some to suggest that individual cells are responsible for encoding abstract concepts. A more likely scenario is that each cell is part of a diffusely distributed network containing several million neurons, and which encodes a memory of the celebrity or landmark. Each individual cell probably contributes to millions of other networks, each of which encodes a distinct memory or concept.

Dendrite: From the Greek word *dendron*, meaning 'tree', this is a branched projection that emanates from the cell body. Dendrites are the neuron's 'input' compartments – they receive and compute signals from other neurons before conveying them to the cell body.

Cell body: This computes different types of signals received from the dendrites and produces an output. It also contains the nucleus, which is packed with DNA – a long molecule containing the information needed to synthesize the thousands of proteins that control the function of the cell. Each type of neuron expresses a unique combination of genes that gives it unique characteristics.

Axon: The axon is a single fibre that emanates from another region of the neuron, and is the 'output' compartment of the neuron. Electrical signals are generated at the initial segment of the axon, and are carried away from the cell body before being transmitted to other cells. The end of the axon forms the nerve terminal, which forms branches that transmit the output of the neuron to multiple 'target' cells. We now know, however, that impulses can be generated in any part of the neuron, and can travel in both directions.

A WELL-DRILLED ORGANIZATION

The vast majority of neurons – about 80 per cent – are found in the cerebellum. The cells in its cortex (the outer covering) are arranged in highly ordered layers, like highly disciplined soldiers in specialist regiments. Two types of cells in this part of the brain illustrate just how diverse these neurons are. Purkinje neurons are the largest cell type in the brain. They are broad, flat and extremely elaborate. Granule cells, by contrast, are the smallest cells in the brain. They have a single fibre that splits in two soon after leaving the cell body, and runs perpendicular to the Purkinje cell dendrites. Each Purkinje cell forms connections with approximately 250,000 granule cell fibres.

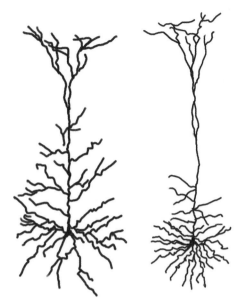

Pyramidal neurons from different parts of the cerebral cortex

The cerebral cortex is also composed of layers, each containing well-ordered neurons. Pyramidal cells, found in all but the outermost layer, are one of the main cell types, and are arranged in clusters in a regular pattern that repeats itself every thirty thousandths of a millimetre. Their structure varies between layers and different brain regions, but all have a distinctive pyramid-shaped cell body, extensively branched dendrites, and a branched axon that extends to cells in other layers of the cortex and distant regions of the brain.

The condensed idea
Neurons are the fundamental components of the nervous system

03 Glial cells

Along with neurons, the brain also contains other cells called glia. For much of the history of modern neuroscience, glial cells were dismissed as nothing more than supporting cells. It's now known that, although glia do assume important supportive roles, they are in fact key players when it comes to brain development, function and disease.

For more than 150 years, glial cells were regarded as necessary merely to hold neurons in place and protect and nourish them. Modern research, though, shows that they also make important contributions to the brain's information-processing capabilities.

In the brain, glial cells outnumber neurons, but ever since their discovery they have largely been neglected by researchers. But it is now becoming increasingly clear that they need to be taken into account if our understanding of how the brain works is to progress. Far from merely being supporting actors, glial cells play important roles on the stage that is brain function, and may yet emerge as the real stars of the show.

GETTING TO KNOW THE GLIAL CELLS

The brain contains different types of glial cells, each of which performs distinct functions:

Astrocytes are star-shaped cells packed into the spaces around neurons. They provide neurons with nutrients and regulate their chemical composition, but are also vital for information processing.

TIMELINE

1839	1856	1896
Theodor Schwann describes the structure of peripheral nerves and observes Schwann cells	Rudolph Virchow names glial cells *nervenkitt*, meaning 'nerve glue'	Gheorghe Marinescu recognizes that glia devour neurons by phagocytosis

Ependymal cells line the walls of the brain ventricles and produce and secrete cerebrospinal fluid. They have hair-like protuberances called cilia that project into the ventricles and beat to aid circulation of the cerebrospinal fluid.

Microglial cells are the brain's emergency response unit, forming a first line of defence against microbes and cleaning up debris from dying neurons (*see pages 126–7*).

Oligodendrocytes produce a fatty tissue called myelin that wraps itself round the axons, enabling nervous impulses to travel along them more efficiently. (Schwann cells perform the same function in the peripheral nervous system.)

Radial glia are present only during early brain development (see *overleaf*). They produce the vast numbers of neurons in the brain and guide their migration into the developing cerebral cortex.

The brain's emergency workers

Microglia are formed in the bone marrow and are the immune cells of the brain. They continuously patrol the brain, extending and retracting their finger-like projections to detect signs of infection, injury or disease. When microglial cells detect microbes that have invaded the brain, they crawl, amoeba-like, towards the invaders, and engulf them by a process called phagocytosis (literally, 'cell eating'). During this process, the microglial cell uses its cell membrane to form an envelope around the microbe, before internalizing the invader then destroying it. Microglia are also deployed when the brain is injured. They detect a chemical distress signal sent out by damaged and dying neurons and respond to it by crawling towards the site of injury. When they arrive at the injury site, they clean up dead cells and other cellular debris.

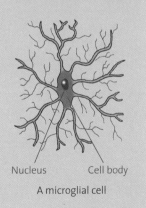

Nucleus   Cell body

A microglial cell

1920	1966	1970
Pio del Rio-Hortega classifies glia into four different types	Stephen Kuffler and colleagues show that glia respond to signals from neurons	Pasko Rakic describes the migration of young neurons along radial glial fibres

MORE THAN JUST GLUE

Translated from the Greek, the word *glia* means 'glue' – reflecting the role that these cells have always been thought to play. But research published in the past ten years shows that glial cells are, in fact, vital for all aspects of brain function.

Astrocytes, for example, are far more than just the packaging that holds neurons in place. They form functioning networks and communicate with each other and with neurons using chemical signals, adding another layer of complexity to the mechanisms of information processing. They also make important contributions to the formation of synapses (connections between neurons) during brain development.

These star-shaped cells control how neurons communicate with each other and are therefore critical for how synapses function in the mature brain. They come into close contact with synapses, clasping them with finger-like protuberances, which can tighten or loosen their grip on the synapses to regulate the flow of chemical signals that pass between neurons. Similarly, astrocytes have other protruberances called endfeet that wrap themselves around capillaries to control blood flow through the brain.

GLIAL CELLS ARE CRITICAL PARTICIPANTS IN EVERY MAJOR ASPECT OF BRAIN DEVELOPMENT, FUNCTION AND DISEASE.

American neurobiologist
Ben Barres, 2008

Astrocytes also regulate something called synaptic plasticity – the process by which neural connections become stronger or weaker in response to experience. These newly discovered functions have led some researchers to suggest that the once-humble astrocytes are critical for mental functions such as memory.

It doesn't end there. Radial glial cells play a critical role in brain development. During the early stages, the nervous system consists of a hollow tube, which will go on to form the brain at one end and the spinal cord at the other. Radial glia have a single fibre that spans the thickness of the tube, and they divide near the inner surface to produce immature neurons.

These young neurons then climb onto the fibres of the radial glial cell that produced them, before crawling along it towards the outer surface of the tube.

This 'radial migration' occurs in waves to produce the characteristic layers of the cerebral cortex, which form in an 'inside-out' fashion – the first wave of migrating neurons form the innermost layers of the cortex, and each subsequent wave migrates past the one before it to produce another layer closer to the outer surface of the tube.

CLEAN-UP FAILURE

Glial cells play a role in many neurological disorders. Multiple sclerosis, for example, is a condition in which the immune system mistakenly attacks oligodendrocytes, breaking down the insulating myelin sheath they produce. This affects the nerves' ability to conduct impulses and causes the symptoms of the disease. In the worst cases, damage to myelin insulating the peripheral nerves causes paralysis, while damage to the optic nerve causes blindness.

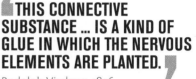

THIS CONNECTIVE SUBSTANCE ... IS A KIND OF GLUE IN WHICH THE NERVOUS ELEMENTS ARE PLANTED.
Rudolph Virchow, 1856

Glial cells are also involved in neurodegenerative conditions such as Alzheimer's disease and Parkinson's. All of these diseases are characterized by abnormally folded proteins that are deposited in or around neurons as insoluble clumps. Normally, microglial cells patrol the brain and clean up any debris, but new evidence suggests that they fail to clear away the protein clumps that accumulate in the brains of patients with neurodegenerative diseases. More recently, researchers have discovered that in people with a condition called amyotrophic lateral sclerosis, a type of motor neuron disease, mutant astrocytes release toxic signals that kill motor neurons.

The condensed idea
Glial cells play key roles in brain function

04 The nervous impulse

Neurons are specialized to produce electrical signals that travel along their fibres. These signals, called nervous impulses or 'action potentials', are generated by the flow of tiny electrical currents across the nerve cell membrane. Neurons can produce up to a thousand action potentials per second, and information is encoded in the pattern of impulses produced.

N ervous impulses are electrical signals that travel along nerve fibres, enabling neurons to communicate with each other and with the rest of the body. The electrical properties of neurons are determined in the cell membrane, which consists of two layers separated by a small gap. The membrane acts as a capacitor, storing electrical charge in the form of ions (positively or negatively charged atoms) and a resistor, which blocks the flow of currents. When a neuron is resting, there is a cloud of negatively charged ions at the inner surface of the membrane and a cloud of positively charged ions outside. This 'charge distribution' makes the inside of the membrane negatively charged with respect to the outside.

When a neuron becomes active, it is said to fire, spike or generate a nervous impulse. This occurs in response to signals received from other cells, and involves a brief reversal of the membrane voltage, with the inside becoming momentarily positively charged before quickly reverting to its resting state. During a nervous impulse, the nerve cell membrane allows in certain types

TIMELINE

1791	1848	1850	1878
Luigi Galvani studies bioelectricity in frogs' legs	Emil du Bois-Reymond discovers the nervous impulse	Hermann von Helmholtz measures the conduction velocity of impulses in frog nerves	Louis-Antoine Ranvier describes nodes in the myelin sheath

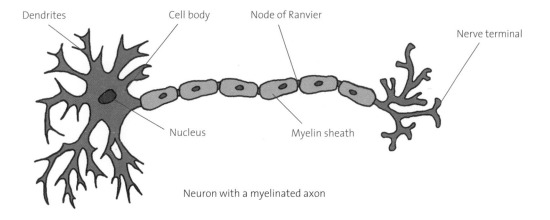

Dendrites   Cell body   Node of Ranvier

Nerve terminal

Nucleus   Myelin sheath

Neuron with a myelinated axon

of ions, which flow back and forth across the membrane. Because ions are electrically charged, their movements constitute a flow of current across the membrane.

NEURONS AT REST

Neurons contain a solution of ions (or electrically charged atoms), and are bathed in a solution that contains the same ions at different concentrations. Ions tend to move from an area where they are highly concentrated to one of low concentration, to produce equilibrium, but are prevented from doing so, however, because the nerve cell membrane is largely impermeable to them.

This creates a situation in which certain ions gather on the outer surface of the membrane, and others gather on the inner surface. The uneven distribution of electrical charges causes the inside of the membrane to be negatively charged, and the outside to be positively charged. The membrane is therefore said to be *polarized*.

IT SEEMS A PLAUSIBLE CONJECTURE ... THAT THE NERVE PERFORMS THE FUNCTION OF A CONDUCTOR.
Luigi Galvani, 1791

1893
Paul Flechsig describes the development of myelin sheaths in the brain

1952
Alan Hodgkin and Andrew Huxley describe the mechanism of the nervous impulse in giant squid axons

1998
Rod MacKinnon and colleagues determine the structure of the voltage-gated potassium channel

Ohm's Law

Ohm's Law explains how the electrical properties of brain cells change in response to incoming signals. It describes the relationship between the voltage of the nerve cell membrane, its resistance and the current that flows across it. It states that the current is directly proportional to the membrane voltage, and is expressed by the equation $I = V/R$, where I is current, V is the membrane voltage, and R is the resistance.

IT STARTED WITH A SQUID

The mechanism of the action potential was determined in the early 1950s, in classic experiments that used microelectrodes impaled in the giant axons of squid. These experiments showed that the action potential is generated by the successive movements of ions across the membrane.

In the first phase of the action potential, the membrane becomes briefly permeable to sodium ions, which flood into the cell. This causes *depolarization* of the cell – the membrane voltage is reversed, becoming positively charged on the inside. This is rapidly followed by a flow of potassium ions out of the cell, which reverses the membrane voltage once again. The influx of potassium ions drives the membrane voltage to become more negative than its resting state, and the cell is said to be *hyperpolarized*. During this so-called refractory period, the neuron cannot produce another action potential, but the neuron quickly reverts to its resting state.

Action potentials are generated at a structure called the axon hillock, the point where the axon arises from the cell body. They travel along the axon because depolarization of one segment of the fibre causes the adjacent area to depolarize as well. This wave of depolarization travels away from the cell body, and when the action potential reaches the nerve terminal, it causes the release of neurotransmitters.

THE MEMBRANE ACTS AS A BARRIER AND PREVENTS THE IONS IN THE EXTERNAL SOLUTION FROM MIXING WITH THE INTERNAL SOLUTION.
Alan Hodgkin, 1964

A single impulse lasts about one-thousandth of a second, and neurons encode information in precisely timed sequences of impulses called spike trains, but exactly how spike trains encode information is still unclear. Neurons often produce action potentials in response to signals from other cells, but they also produce impulses in the absence of signals. The basal firing rate, or frequency of spontaneous action potentials, varies between different types of neurons, and can be altered by signals from other cells.

Faster than Usain Bolt

Axons in the brain and spinal cord are insulated by fatty myelin tissue produced by brain cells called oligodendrocytes. An oligodendrocyte has a small number of branches, each of which consists of a large, flat sheet of myelin that wraps itself many times around a small segment of an axon belonging to a different neuron. The myelin sheath is not continuous along the length of an axon, but is interrupted at regular intervals by gaps called Nodes of Ranvier. Ion channels are clustered at the nodes, allowing action potentials to jump from one node to the next. This process increases the velocity at which they are propagated along the axon, anything up to 100m (328ft) per second.

FEW SHALL PASS

Ions flow across the nerve cell membrane through barrel-shaped proteins called ion channels, which are embedded in the membrane and form pores through it. Ion channels contain sensors that detect changes in the membrane voltage, and open and close in response to these changes.

Human neurons contain more than a dozen different types of ion channel, each of which allows only one species of ion to pass through it. The activity of all these ion channels is tightly orchestrated during the action potential. They open and close in sequence, so that neurons can generate patterns of nervous impulses in response to signals received from other cells.

The condensed idea
Neurons produce electrical signals that carry information

05 Synaptic transmission

Nerve cells communicate with each other through a process called neurochemical transmission. This occurs at junctions called synapses, and involves chemicals called neurotransmitters, which pass between adjacent neurons and carry signals. Neurochemical transmission is modified by learning, and drugs work by altering it in one way or another.

Electrical signals produced by the neurons cannot simply jump between cells, so they are converted to chemical signals that can be transmitted from one cell to the next. This process, called neurotransmission, takes place at highly specialized junctions called synapses, and involves chemicals called neurotransmitters, which pass between cells. As a general rule, neurons synthesize and release one type of neurotransmitter; they form precise connections, so that each type of signal is targeted to specific 'target' cells. Learning and memory are thought to involve the modification of synapses within networks of neurons, and drugs exert their effects by somehow altering synaptic transmission.

THE SUPER SYNAPSE

Synapses consist of two specialized components: the *presynaptic terminal* of the cell producing the signal, and the *postsynaptic cell* that receives the signal. At the presynaptic terminal, neurotransmitter molecules are stored in tiny spherical structures called synaptic vesicles, which are 'docked' at the so-called active zone near the cell membrane. When an action potential arrives at the

TIMELINE

1897	1914
Charles Sherrington coins the term *synapse*, meaning 'to clasp'	Henry Dale and colleagues identify acetylcholine as a potential neurotransmitter

nerve terminal, it causes the vesicles to fuse with the membrane and release their contents into the synapse.

Once released, neurotransmitter molecules diffuse across the synapse and bind to receptors embedded in the membrane of the postsynaptic neuron. Some receptors alter the electrical properties of the post-synaptic cell directly, by allowing small currents to flow in or out of the cell. Others do so indirectly and more slowly, by initiating biochemical signalling pathways. After being released, neurotransmitters are usually mopped up by the cells that released them, a process called reuptake.

Mind the gap

Neurons also communicate with each other via electrical synapses called gap junctions. These are made of proteins called connexins, which span the cell membranes and connect the inside of adjacent cells. Gap junctions allow for instantaneous electrical signalling between neurons, so that networks of interconnected cells can fire in synchrony as currents pass between them.

Neurotransmission is a complex process involving the orchestrated actions of hundreds of proteins on both sides of the synapse, each performing a specific function. In presynaptic neurons, dozens of proteins participate to control the fusion of synaptic vesicles to the presynaptic membrane. On the opposite side of the synapse, dozens of receptors and numerous other components of the signalling machinery are arranged in a highly organized manner so that the signals can be put into effect efficiently. As the brain processes information, it alters synapses by modifying the efficacy of neurotransmission in various ways. The number of docked vesicles can be increased or decreased, so that more or fewer neurotransmitter molecules are released. On the other side of the synapse, receptors can be inserted or removed from the postsynaptic membrane, to make the cell more or less sensitive to the signals it receives.

> IN VIEW OF THE PROBABLE IMPORTANCE OF THE ... NEXUS BETWEEN NEURON AND NEURON IT IS CONVENIENT TO HAVE A NAME FOR IT. THE TERM INTRODUCED HAS BEEN *SYNAPSE.*
>
> Charles Sherrington, 1906

1921

Otto Loewi provides the first evidence of synaptic transmission

1936

Dale and Loewi share the Nobel Prize in Physiology for their work on acetylcholine

The dream of Otto Loewi

Neurotransmission was discovered in 1921 by Otto Loewi, in an experiment that apparently came to him in a dream. Loewi dissected two frogs' hearts, one with the vagus nerve still attached to it. He placed each heart in a separate container filled with salty water, and electrically stimulated the vagus nerve, causing the heart to which it was attached to slow down. Loewi then transferred some of the salt-water solution from that container into the other, and found that the second heart slowed down, too. The experiment confirmed that electrical stimulation causes the vagus nerve to release a chemical signal that slows the heart rate. Loewi named the chemical *vagusstoff*, meaning 'vagus substance', but it was soon identified as acetylcholine, which had been discovered a few years earlier by Henry Dale.

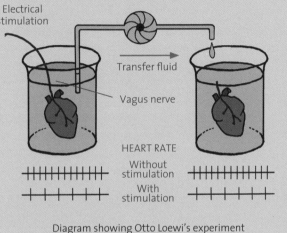

Diagram showing Otto Loewi's experiment

WHY WE NEED NEUROTRANSMITTERS

The brain contains something like one quadrillion (one thousand million million) synapses, and produces about one hundred different neurotransmitters. Glutamate, gamma-aminobutyric acid (GABA) and glycine are amino acid transmitters. The monoamines are another group of transmitters, including dopamine, adrenaline and serotonin. Dopamine is often referred to as 'the pleasure molecule' because it is involved in reward, but it also has important roles in attention, memory and movement. Serotonin plays a vital part in mood.

The neuropeptides are small proteins crucial in pain signalling, while the endocannabinoids, a group of transmitters that have gained increasing attention in recent years, are involved in appetite, mood and memory. Other neurotransmitters include acetylcholine – which motor neurons use to send signals to the muscles, and which is also used in the autonomic nervous system – and nitric oxide, a gas thought to play an important role in learning and memory.

EXCITATION VERSUS INHIBITION

Neurotransmitters can be broadly divided into two different types according to their effects on neurons: excitatory neurotransmitters depolarize the nerve cell membrane, making the cell more likely to generate action potentials, whereas inhibitory neurotransmitters make the membrane voltage more negative, so that the cell is less likely to fire (*see page 16*).

Proper brain function depends on a delicate balance between excitation and inhibition, and disturbing this balance can have dramatic effects. Epilepsy, for example, is characterized by seizures thought to occur due to too much excitatory neurotransmission.

HOW DRUGS WORK

Some drugs have molecular structures similar to neurotransmitters, and therefore mimic their actions. LSD, for example, resembles serotonin, and activates serotonin receptors by binding to them in place of the neurotransmitter. Other drugs activate neurotransmitter receptors by binding to specialized sites. The GABAA receptor, for instance, contains a region that binds to the anti-anxiety drug diazepam and related compounds. These drugs therefore alleviate anxiety by activating GABAA receptors in certain brain regions, which enhances inhibitory synaptic transmission. Yet others act by enhancing or blocking neurotransmitter reuptake. Prozac and related antidepressants, for example, are referred to as selective serotonin reuptake inhibitors (SSRIs). They prevent neurons from reabsorbing serotonin after neurotransmission has taken place, thus prolonging the effects of this transmitter at synapses.

The condensed idea
Neurons communicate with each other through chemical signals

06 Sensory perception

The senses are the windows through which information about the outside world enters the brain. Each sense organ is specialized to detect data in the form of physical energy, which is then converted into electrical impulses that are sent to the brain, where they are processed and interpreted to generate a coherent experience of the world.

The brain evolved to detect and respond to changes in the environment, and receives information about the world through the sense organs. Each sense organ detects a specific type of sensory stimulus, which it translates into the electrochemical language of the brain. The five senses – vision, hearing, touch, taste and smell – were recognized more than 2,000 years ago by the Greek philosopher Aristotle. The scientific study of perception began in the 19th century, and modern neuroscience provides a deeper understanding of the mechanisms involved.

All of the brain's sensory systems share a common ground plan. The first stage of perception is called sensory transduction – it is the process by which specialized receptors detect physical stimuli from the environment and convert them into electrical impulses. The information is then sent to a part of the brain called the thalamus (or 'deep chamber'), which relays it to the appropriate region of the cerebral cortex.

INSIGHT INTO SIGHT
Vision is the best understood of all the senses. The retina contains several different types of photoreceptors, which are sensitive to light particles called photons. When light strikes the retina, it causes biochemical reactions in

the photoreceptors. The photoreceptors then transmit signals carrying the light information to other cells in the retina, which perform the early stages of vision processing. The information is then transmitted, via the optic nerve, to a part of the thalamus called the lateral geniculate nucleus, which relays it to the visual cortex.

The visual cortex is located at the back of the brain in the occipital lobe, and contains dozens of distinct subregions. Each is specialized to perform a particular function, and visual information is processed in a hierarchical manner. The visual cortex contains multiple pathways, each of which processes a different type of information. These pathways process data in parallel, and then converge at the final stages of processing.

Processing begins in the primary visual cortex (also called Area V1), which contains cells that respond to basic features of an image, such as contrast and the orientation of edges. Information passes from one subregion to the next, becoming increasingly complex at each successive stage. Thus, the basic features of an image – such as shape, colour and movement – are woven together as they progress through the visual pathway, so that the pattern of light that fell upon the retina is reconstructed into the dynamic image of the world that we 'see'.

HEAR THIS

The ear channels sound waves onto the ear drum, which transmits them to the cochlea, a spiral-shaped structure containing three fluid-filled cavities. Sound waves cause the fluid to move around, and the movements are detected by receptors called hair cells, each of which is sensitive to sound waves of a specific frequency.

The 'sixth' sense

Proprioception, sometimes called the 'sixth' sense, refers to our sense of limb position and movements. Muscles contain stretch receptors called muscle spindles, which detect changes in muscle length and convey them, via peripheral nerves, into the spinal cord. The signals are then transmitted to the brain, which uses them to generate a postural model of the body.

1880	1911	1916	2004
Francis Galton describes grapheme-colour synaesthesia	Allvar Gullstrand wins the Nobel Prize for his work on optics	Shinobu Ishihara publishes colour blindness test	Linda Buck and Richard Axel win the Nobel Prize for their work on smell

Joined-up senses

Synaesthesia means 'joined senses', and refers to a phenomenon in which stimulation of one sense produces sensations in another. Physicist Richard Feynman was what is known as a 'grapheme-colour synaesthete' – in other words, he experienced specific colours in response to letters and numbers – while expressionist artist Wassily Kandinsky was a 'tone-colour synaesthete', who associated musical tones with colours. Other forms include 'mirror-touch synaesthesia' (seeing another person being touched evokes touch sensations), and 'time-space synaesthesia' (units of time, such as days and months, are seen as occupying specific locations in space relative to the body). Synaesthesia was once thought to be extremely rare, but is now estimated to affect about 1 per cent of the population. According to one theory, it arises when connections between different sensory pathways that are normally eliminated during brain development remain in place. Another states that it occurs because of too much 'cross-talk' between sensory pathways.

Information is carried by the auditory nerve, via the thalamus, to the temporal lobes of the brain, which contain areas specialized for processing sound. The temporal lobe contains language areas; if these are damaged, it can cause difficulties in producing or understanding speech. The auditory nerve also sends the information to the inferior colliculus, a part of the brain stem that determines where the sound is coming from by comparing the signals from the left and right ear.

SENSING THE WORLD

The somatosensory system processes touch, pain and temperature, which are detected by receptors in the nerve endings near the skin surface. This information is carried along the peripheral nerves into the spinal cord and then up to the brain, where it is processed in the primary somatosensory cortex. The cells involved each have a single fibre that extends from just beneath the skin surface into the spinal cord – they are the longest cells in the nervous system.

The nerve endings of these sensory neurons contain numerous receptors specialized to detect different types of somatosensory information. Some receptors detect cold or hot temperatures, for example, while others detect touch, itch or pain. Each type of information is carried into the spinal cord by dedicated nerve fibres.

Pain information is transmitted by special sensory neurons called noci-ceptors, which contain receptors that detect one or more noxious stimuli, such as extremely cold or hot temperatures, excessive mechanical pressure or dangerous chemicals. They also contain receptors sensitive to various chemicals that are released by damaged cells.

THE SCIENCE OF SMELL AND TASTE

The inside of the nose is lined with a thin sheet of tissue containing around 1,000 different types of olfactory receptors, which detect airborne odorant molecules. The cells containing these receptors give rise to axons that project to several different parts of the brain, which together allow the perception of smells and the social cues they entail. Chemicals called pheromones play important roles in animal and probably human behaviour.

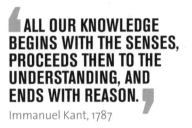

ALL OUR KNOWLEDGE BEGINS WITH THE SENSES, PROCEEDS THEN TO THE UNDERSTANDING, AND ENDS WITH REASON.
Immanuel Kant, 1787

In the tongue, taste buds contain receptors that detect salty, sour, bitter and sweet tastes, and a savoury sensation called umami. New research shows that taste preferences are at least partly determined by genetics. For example, variations in the gene encoding the OR7D4 olfactory receptor determine sensitivity to androstenone, a pheromone found in cooked pig meat, and people carrying two copies of a particular variant rate pork meat less favourably than do others. Taste and smell are the least understood of the human senses, but we know that they are intimately linked. To see just how closely, hold your nose while eating something, and you will find that you cannot taste the food.

The condensed idea
The brain receives 'inputs' through the sense organs

07 Movement

Movement is one of the main functions of the nervous system, and a large part of its work is devoted to the planning and execution of movement. Moving involves multiple parts of the brain as well as the spinal cord, working together to control the body muscles. The motor system is damaged in Parkinson's disease and other movement disorders.

Generating movement is one of the primary functions of the nervous system. All animals – including human beings – must move to find mates, locate food, evade predators and escape from potentially dangerous situations. Consequently, a great deal of the brain is devoted to planning and executing voluntary movements. The motor system of the human brain includes parts of the cerebral cortex as well as various subcortical structures and the spinal cord. These structures work together to make movement possible.

Cerebral cortex: The frontal lobe contains several distinct areas that are specialized for movement. One is the supplementary motor cortex, which is involved in planning movement. Another is the premotor area, which encodes the intention to perform a particular movement, and selects the appropriate movement based on sensory information. The primary motor cortex at the back of the frontal lobe contains large neurons called Betz cells, which send long fibres down into the spinal cord, where they form synapses with the motor neurons that send signals to the muscles. These fibres cross from one side of the nervous system to the other as they pass through the brain stem. Thus, each hemisphere of the brain controls the movements of the opposite side of the body.

TIMELINE

1800	1817	1823
Samuel von Sömmerring identifies the substantia nigra	Publication of *An Essay on the Shaking Palsy* by James Parkinson	Marie-Jean-Pierre Flourens discovers the role of the cerebellum in regulating movement

Knee-jerk reactions

The spinal cord contains neuronal circuits that can initiate simple, involuntary movements without your brain having to be involved. One example is the knee-jerk reflex, which is used by doctors to test for spinal injuries. This reflex involves a simple spinal circuit containing just two neurons. Striking the bottom of the kneecap with a hammer activates a stretch receptor in the quadriceps muscle, which sends a nervous impulse along the nerve fibre of a sensory neuron into the spinal cord. In the spinal cord, the sensory fibre forms a synapse with a motor neuron. The impulse is transmitted to the motor neuron, which then transmits a signal to the leg muscle, causing the lower leg to jerk forward. The knee-jerk reflex takes about 50 milliseconds, and is referred to as 'monosynaptic' because it involves just one synapse. It normally helps us to maintain proper posture.

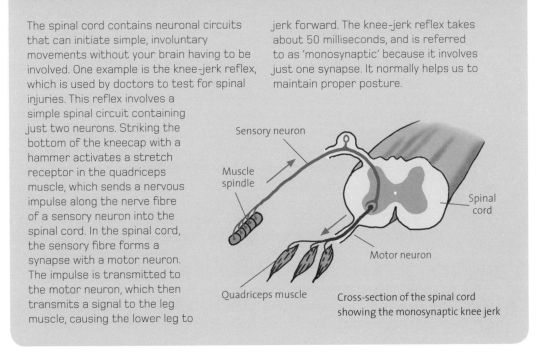

Cross-section of the spinal cord showing the monosynaptic knee jerk

Basal ganglia: The basal ganglia are a group of subcortical structures located beneath the frontal cortex, which together make up the corpus striatum (or 'striped body'). They are involved in a variety of functions, among them the control of voluntary movement, and they connect almost exclusively with the cerebral cortex. According to one hypothesis, the basal ganglia generate various patterns of movements that are executed by the cerebral cortex. They then receive feedback about the results of each pattern, and reinforce the most successful one with a rewarding dopamine signal. Recent research suggests,

1874

Vladimir Alekseyevich Betz discovers giant pyramidal neurons in the primary motor cortex

1924

Charles Sherrington discovers the stretch reflex

however, that they learn new skills rapidly, by monitoring the variations in movements then training the cortex to execute the most appropriate option. The basal ganglia are affected in movement disorders such as Parkinson's and Huntingdon's diseases.

Cerebellum: The cerebellum (or 'little brain') lies behind the brain stem and is involved in balance and the control and coordination of movement. It does this by integrating sensory signals with information from the motor areas of the cerebral cortex. The cerebellum also plays important roles in the timing and precision of movements, and also in motor learning. Learning a motor skill initially requires a lot of attention, but once the skill has been learned, the movements involved can be executed effortlessly, and almost subconsciously, largely because they have been programmed into the wiring of the cerebellum. The connections between the two main cell types in the cerebellum – the Purkinje cells and granule cells – continue to form well after birth; this is why it takes time for young children to learn how to walk and to perform fine motor skills. And the effects of alcohol on the cerebellum explain why people stagger when drunk.

WE HAVE A BRAIN FOR ONE REASON AND ONE REASON ONLY, TO PRODUCE ADAPTABLE AND COMPLEX MOVEMENTS.

British neuroscientist
Daniel Wolpert, 2011

Spinal cord: The human spinal cord consists of 31 segments, each containing a collection of motor neurons whose fibres extend to the muscles of the body. Spinal motor neurons are located towards the back of the spinal cord; their fibres emerge from the back of the cord in the spaces between the vertebrae, and are bundled together with sensory nerve fibres, which emerge from the front of the spinal cord to form the peripheral nerves. Motor neurons in the spinal cord receive signals from motor neurons in the primary motor cortex, and in turn signal the muscles to control their contractions. Motor neurons communicate with muscles at a specialized synapse called the neuromuscular junction; an individual motor neuron and the muscles it controls are collectively referred to as a motor unit. Voluntary movements are planned by the brain, and the instructions to execute them are then sent down to the spinal cord, but the spinal cord can initiate simple movements called reflexes without the involvement of the brain (*see page 29*).

MOVEMENT DISORDERS

A number of neurodegenerative disorders affect our ability to move, all of which involve damage to some component of the brain's motor system. Parkinson's disease, for example, is characterized by the death of dopamine-producing neurons in part of the basal ganglia called the substantia nigra, leading to tremor, muscle rigidity and bradykinesia (slow movements).

In Huntington's disease, on the other hand, the striatum is damaged during the early stages of the disease, leading to uncontrollable movements that become more frequent and extreme as the disorder progresses. (The condition was originally named Huntington's chorea, with the term *chorea* derived from the Greek word meaning dance, to reflect these symptoms.)

Motor neuron diseases are another group of disorders that affect movement. As the name suggests, these conditions involve the death of 'upper' motor neurons in the cerebral cortex or the 'lower' motor neurons in the spinal cord. This causes problems with walking, speaking, breathing and swallowing, leading to progressive disability and, eventually, death.

Strokes can affect movement, too. They often cause damage in or around the motor cortex of the left hemisphere, including regions that control the muscles needed to produce speech, leading to the classic symptoms of paralysis on the right side of the body and an inability to speak.

The condensed idea
Movements are the main output of the brain

08 Topographic mapping

The body's surface and certain features of the external world are mapped onto the brain in a highly ordered fashion. These so-called 'topographic maps' exist in all of the brain's sensory systems, as well as in its motor system. They arise during brain development, and are vital for information processing.

In the 1920s, the neurosurgeon Wilder Penfield pioneered a technique for electrically stimulating the brains of conscious patients, in order to locate and remove abnormal brain tissue causing seizures, while sparing surrounding tissue involved in important functions such as language and memory. To do so, he applied local anaesthetic to the patients' scalps, opened up their skulls to expose the surface of the brain, then used electrodes to apply a current.

Penfield operated on about 400 patients altogether in this way, using his electrodes to probe many brain regions systematically. Because his patients remained conscious throughout, they could report back on the sensations they experienced. He found that stimulating the medial temporal lobe, for example, could evoke vivid memories, while stimulation of parts of the visual cortex caused his patients to perceive simple patterns of light. But his most important and best-known discovery was that the primary motor and somatosensory cortices contained maps of the body.

When Penfield stimulated a part of the left somatosensory cortex, a strip of tissue that runs down the front of the parietal lobe, the patient might report

TIMELINE

1928	1943
Wilder Penfield develops his technique for electrical brain stimulation in conscious surgical patients	Roger Sperry proposes the chemoaffinity hypothesis following experiments on the African clawed toad

a tingling sensation in their right hand. Moving the electrode a centimetre farther up the somatosensory strip would evoke touch sensations in the forearm or elbow. A similar situation exists in the motor cortex, which lies in the frontal lobe. Stimulating a part of the motor cortex evoked muscle twitches or small movements on the opposite side of the body, and moving the electrode to an adjacent area evoked the same response in an adjacent body part.

Penfield found that, although there were small differences between individuals due to variations in brain structure, the overall organization of these maps was basically the same in all of his patients: the body was represented on the surface of the brain in a highly ordered fashion, with adjacent body parts mapping precisely onto adjacent areas of the brain. Penfield's pioneering electrical stimulation technique is still used by neurosurgeons today, albeit with more advanced technologies. And his discoveries about the topographic organization of the motor and somatosensory cortices, illustrated by the famous 'homunculus' (*see page 34*), are still highly relevant.

> **GROWING FIBERS ARE EXTREMELY PARTICULAR WHEN IT COMES TO ESTABLISHING SYNAPTIC CONNECTIONS, EACH AXON LINKING ONLY WITH CERTAIN NEURONS TO WHICH IT BECOMES SELECTIVELY ATTACHED BY SPECIFIC CHEMICAL AFFINITIES.**
>
> Roger Sperry, 1963

THE VISUAL MAP

Visual information entering the eye is likewise mapped onto the brain, and this phenomenon is referred to as retinotopy. During the first stage of visual processing, light energy falls onto the photoreceptors at the back of the retina, with adjacent locations of the visual field falling onto adjacent areas of the retina. This topographic organization is maintained throughout the visual system. The optic nerve exits the back of the eye and projects to a part of the thalamus called the lateral geniculate nucleus (LGN), which then relays visual information to the visual areas in the occipital lobe at the back of the brain. Adjacent retinal cells send fibres to adjacent regions of the LGN, which in turn project to adjacent regions of the primary visual cortex.

1950	2012
Wilder Penfield summarizes his life's work in the book *The Cerebral Cortex of Man*	Researchers discover 'tunotopic' organization of the mouse olfactory system

Little men in the brain

The neurosurgeon Wilder Penfield mapped the body onto the brain by electrically stimulating the cortex of conscious epileptic patients and noting their responses. He found that some body parts are represented by disproportionately large areas of the motor and somatosensory cortices. The size of the cortical representation of a body part depends on the number of nerve endings it contains. The hands and face are the most sensitive parts of the body, and also contain more muscles than any other part. Therefore, between them they take up the vast majority of the primary motor and somatosensory cortices, as depicted in the now famous homunculi, which were originally drawn by Penfield's secretary. Penfield also found that the feet are represented next to the genitals in the somatosensory cortex, and used this to explain why some people have foot fetishes. Recent work has failed to confirm these findings, however.

The famous three-dimensional homunculus sculpture

The leading hypothesis for how retinotopic maps are formed comes from a series of classic, if rather gruesome, experiments performed on the African clawed toad in the 1940s. Roger Sperry severed the toads' optic nerves, rotated their eyes by 180°, then replaced them. Over a period of several weeks, the optic nerve fibres regenerated and grew back to the tectum, the main visual processing region of the amphibian brain. When he tested their vision, however, he found that it was inverted. When he dangled a fly above them, they would extend their tongues downwards. When the fly was shown on their right, they would extend their tongue to the left.

These findings showed that the regenerated optic nerve fibres somehow find their way back to their original destinations in the tectum. Sperry explained this with the 'chemoaffinity hypothesis', according to which the optic nerve fibres and their targets in the tectum possess complementary molecular 'tags' with which they find each other. This is borne out in modern research, which shows that growing nerve fibres are indeed guided along the right path by specific chemical signals.

SOUND AND SMELL MAPS

Topographic maps also exist in ear and brain structures involved in hearing. The cochlea, a spiral-shaped structure in the inner ear, contains cells sensitive to sound waves of different frequencies. Normally, we can hear sound waves with frequencies ranging from 20 to 20,000 Hz, and these frequencies are associated with pitch – with lower frequencies corresponding to lower pitched sounds.

Hair cells that respond to the lowest pitch are located at one end of the cochlea, and those responding to the highest pitch are at the other. As is the case in the visual system, this 'tonotopic' arrangement is maintained in the primary auditory cortex at the top of the temporal lobe. Here, neurons are arranged in bands that are tuned to specific ranges of frequencies. The band at the front end contains cells tuned to frequencies of up to 500 Hz, the next one contains cells tuned to frequencies of between 500 and 1,000 Hz, and so on.

The latest research shows that the olfactory system is organized in a similar way. The olfactory bulb contains structures called glomeruli, containing neurons that respond to specific smells. The glomeruli are arranged in clusters according to the odours to which they are tuned – cells that bind to odorant molecules with a similar structure are located next to each other.

The condensed idea
The brain contains maps of the body and external space

09 Specialized brain regions

Does the cerebral cortex consist of dozens of discrete regions, each comprising distinct types of cells and performing a specialized function? This idea has had a major influence on our understanding of brain function, but is opposed by an alternative theory, which states that interconnected brain regions work together.

The idea that cognitive functions are localized to specific parts of the brain is referred to as 'functional modularity', or the 'localization of cerebral function'. Its origins can be traced back to the late 18th century, when Franz Joseph Gall developed phrenology, which linked personality traits to bumps on the skull. Phrenology was hugely popular throughout the 19th century, but was eventually dismissed as a pseudoscience.

The localization of cerebral function gained scientific credibility in the mid- to late 19th century through the work of two neurologists who observed that patients with speech deficits had suffered damage to distinct parts of the brain. One of these was the French physician Pierre Paul Broca, who studied patients who had lost the ability to speak after suffering a stroke. Broca examined their brains after they died, and noticed that they had all sustained damage to the same part of the left frontal lobe. This region came to be known as Broca's Area, and the inability to produce speech following a stroke is often referred to as Broca's aphasia.

TIMELINE

1796	1861	1874
Franz Joseph Gall develops phrenology	Pierre Paul Broca presents his work on stroke patients	Carl Wernicke publishes his work on stroke patients who cannot understand speech

About ten years later, a German physician named Carl Wernicke studied stroke patients who had lost the ability to understand spoken language. When he examined their brains, he noticed that they had all sustained damage to another specific part of the brain. Wernicke's Area, as it is now known, is located in the left temporal lobe, and the inability to understand speech after a stroke is sometimes referred to as Wernicke's aphasia. (*See also Chapter 28: Language processing.*)

BRODMANN'S AREAS

In the early part of the 20th century, the German anatomist Korbinian Brodmann systematically analysed and compared the cerebral cortex of human beings, monkeys and various other species of mammals. He dissected the tissues from different parts of the cortex, stained them using a technique called Nissl staining, and examined their fine structure under the microscope. Although the cortex has a uniform, layered structure, Brodmann noticed some subtle differences. In some areas, certain layers were more prominent than others and more densely packed with neurons.

Brodmann also noticed that these differences in cellular organization defined boundaries between adjacent regions. He divided the human cerebral cortex into 43 distinct regions on the basis of these observations, and published a map of it in 1909. Brodmann's map has been used extensively over the course of the past century, and remains pertinent to this day. The primary motor cortex, for example, is often referred to as Brodmann's Area 4, and the primary visual cortex is also known as Area 17.

> **THE GREAT NUMBER OF SPECIALLY PREPARED STRUCTURAL AREAS SUGGESTS A SPECIAL SEPARATION OF INDIVIDUAL FUNCTIONS.**
> Korbinian Brodmann, 1909

1909	1920s	1947	1992
Korbinian Brodmann publishes comparative studies of the cortex	Karl Lashley tries to localize the memory trace in rats' brains	Joachim Bodamer coins the term 'prosopagnosia'	Justine Sergent describes the fusiform face area

Knowing me, knowing you

Faces are particularly important for our social interactions, but there is debate over whether the brain treats them as special stimuli or as a subcategory of objects. The fusiform face area (FFA), located on the lower surface of the temporal lobe, responds highly selectively to faces, suggesting that it is specialized for processing faces, and that faces are indeed special. Damage to the FFA or surrounding areas causes prosopagnosia, or face blindness, a condition characterized by the inability to recognize faces. In extreme cases, someone with prosopagnosia fails to recognize even their own face in the mirror or in photographs. In his book *The Man Who Mistook His Wife for a Hat*, the neurologist Oliver Sacks described the case of a farmer who could recognize the cows that he kept by their facial features, but lost the ability to do so as a result of brain damage.

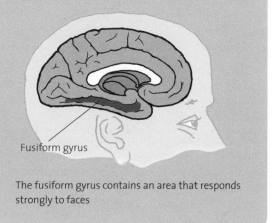

Fusiform gyrus

The fusiform gyrus contains an area that responds strongly to faces

Researchers have confirmed Brodmann's original observations using modern techniques, but they have also revealed more details that refine his original map. For example, Brodmann listed five areas in the monkey brain (areas 17–21) as being devoted to the processing of visual information, but modern anatomical and physiological techniques have revealed that these areas can be further subdivided into around 40 distinct regions, each with its own distinct function.

ALL WORKING TOGETHER

Some researchers have criticized the idea that the cerebral cortex contains discrete, specialized areas and favour instead the idea of 'distributive processing'. One such was the physiologist Karl Lashley. In the 1920s, Lashley performed a series of famous experiments designed to determine where memories are stored in the brain. He taught rats to find their way through a maze, then damaged parts of their cerebral cortex to try to erase the memory trace. He found that the rats could always find their way through the maze again, no matter where the damage was. Based on these observations, Lashley concluded that memory function is not located in a discrete region of the cortex, but is instead distributed throughout the brain.

It has also been observed that some brain areas apparently specialized to perform functions such as vision or sound processing can also undertake other jobs. A study published in 2012, for example, showed that in people who are born deaf, the auditory cortex – which normally processes sound information – can process touch and visual information.

Functional modularity and distributive processing are not mutually exclusive, however. Indeed, the current view of how the brain works is a combination of the two ideas. Neuroscientists now think that the brain operates as what they call a 'massively parallel distributed processor', with multiple networks working together to generate thoughts and behaviour. In other words, the brain does contain discrete areas specialized to perform specific functions, but individual specialized brain areas do not act on their own. Instead, each one can be thought of as a node within a network that is distributed throughout the brain or within particular regions.

Each network contains multiple, interconnected brain regions cooperating to encode particular types of information or generate certain behaviours. Within each network, information is processed serially, or transferred from one area to the next, processing it one after the other. Multiple networks probably work simultaneously, and, at a higher level of organization, activity from multiple networks is integrated, giving rise to the summed patterns of activity that generate our thoughts and behaviours.

The condensed idea
The cortex contains distinct areas for specific functions

10 Brain asymmetry

There are anatomical differences between the left and right hemispheres of the brain, and also a division of labour, with certain functions being localized to one side or the other. Language is largely confined to the left hemisphere, while spatial abilities and perception are dependent largely on the right, and the discovery of these asymmetries has led to the popular left brain/right brain myth.

Brain asymmetries are evident in most animals, and probably emerged in our evolutionary ancestors around 500 million years ago. The human brain is divided into left and right hemispheres, each of which controls the opposite side of the body. The two hemispheres are connected to each other by the corpus callosum (or 'hard body'), a massive bundle of several hundred million nerve fibres, and two smaller fibre tracts, one towards the front and the other towards the back of the brain. At first glance, the left and right hemispheres look like mirror images of each other, but on closer inspection it becomes apparent that they differ in size and shape.

THE SAME BUT DIFFERENT

The left and right hemispheres of the brain are similar in weight and volume, but the left hemisphere is usually slightly larger than the right. The left hemisphere also protrudes slightly at the back, whereas the right hemisphere protrudes from the front. The frontal and central regions, which contain areas

TIMELINE

1861	1868	1874	1884
Broca describes the speech centre in the left frontal lobe	John Hughlings Jackson describes spatial deficits in patients with right-hemisphere damage	Carl Wernicke describes the speech centre in the left temporal lobe	Asymmetries of the Sylvian fissure reported

The left brain/right brain myth

People often say that the left hemisphere of the brain is logical and analytical while the right hemisphere is artistic and creative, and certain patterns of thought and behaviour can be cultivated by harnessing the activity in one side or the other.

The left brain/right brain myth probably originated in the studies of split-brain patients examined during the 1960s, which confirmed earlier observations that speech functions are predominantly performed by the left hemisphere,

and certain perceptual and spatial abilities by the right.

There's little dispute among neuroscientists about these functional asymmetries, but in reality the brain acts as an integrated whole, and most of our behaviours involve both hemispheres acting together in a coordinated manner. The left brain/right brain myth is nevertheless appealing, and is often exploited to market products or services that promise, for example, to tap into your right brain's creative potential.

related to speech and movement, are often wider in the right hemisphere than in the left, and the occipital lobe, which contains areas involved in vision, is often wider on the left than on the right.

Some of the anatomical differences between the two hemispheres are already clearly visible before birth. In most right-handed people, for example, the Sylvian fissure, a prominent groove that separates the temporal lobe from the frontal and parietal lobes, is longer in the left hemisphere than in the right, and also runs at a slightly shallower angle. Differences between the Sylvian fissure and surrounding regions of the left and right hemispheres were among the earliest brain asymmetries to be identified, and are probably related to language functions, which are largely – but not exclusively –

1968
Neurologists report that the planum temporale is larger on the left than the right in more than two-thirds of people

1969
Roger Sperry and Michael Gazzaniga publish initial tests of split-brain patients

1985
Arnold Scheibel and colleagues report asymmetries in dendrites

confined to the left temporal and frontal lobes. Some studies have also noted that the left hemisphere develops more slowly than the right during the first year of life, but eventually takes over, and this, too, is likely related to the development of language functions.

Most studies of brain asymmetry focus on differences in overall size and structure, but anatomical asymmetries are also observable at the microscopic level. In the cortex, cells are organized in regular, repeating columns, and cellular organization in certain brain regions differs between the hemispheres. For example, the columns in the language areas of the left hemisphere are wider than those in corresponding regions on the right. According to one study, the dendrites of cells in these areas of the left hemisphere are longer, and branch more extensively, than those on the right.

LATERAL THINKING

The left and right hemispheres of the brain also differ in the functions they perform. This phenomenon, referred to as the lateralization of cortical function, became clear during the 19th century, largely from studies of brain-damaged patients. During the 1860s and 1870s, Pierre Paul Broca and Carl Wernicke examined stroke patients with speech deficits; when these patients died, postmortem examinations of their brains revealed damage to specific parts of the left hemisphere, which thus came to be associated with language functions.

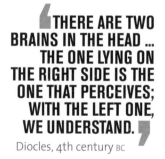

❝THERE ARE TWO BRAINS IN THE HEAD ... THE ONE LYING ON THE RIGHT SIDE IS THE ONE THAT PERCEIVES; WITH THE LEFT ONE, WE UNDERSTAND. ❞

Diocles, 4th century BC

At around the same time, the British neurologist John Hughlings Jackson observed that damage to the right hemisphere often led to impaired perception and spatial abilities. He described one patient who was paralysed on the left side of his body and who had lost the ability to recognize places, objects and people, including his own wife. Another had completely lost her sense of direction, and when she died, a large tumour was found near the back of her right temporal lobe. Based on these observations, Jackson concluded that speech must be mediated by the left hemisphere, and spatial functions by the right.

More recently, studies of split-brain patients have reinforced the idea that brain functions are lateralized. These patients, of which there are only a few, had the corpus callosum severed in order to bring their drug-resistant epilepsy under control, and to prevent the seizures from spreading from one hemisphere to the other. Such patients lead otherwise normal lives, but certain behavioural quirks became apparent in lab tests. For example, they can name and describe objects placed into their right hand, but not the left. When objects are placed in their left hand, the touch information enters the right hemisphere but cannot cross to the speech centres on the left.

MYSTERIOUS BRAIN ASYMMETRY

Brain asymmetry, language and handedness are somehow related, but the relationship between them is complex and very poorly understood. Language functions are strongly lateralized to the left hemisphere in the vast majority (about 97 per cent) of right-handed people, and also in about 70 per cent of left-handers. In a minority of people, speech is strongly represented on the right, or bilaterally (in both hemispheres). Consequently, the left hemisphere of the brain is sometimes said to be 'dominant' over the right.

> **❝DAMAGE TO BUT ONE HEMISPHERE WILL MAKE A MAN SPEECHLESS.❞**
> John Hughlings Jackson, 1874

Animals also exhibit brain asymmetry and handedness, but exactly why brains evolved this way, and why the speech centres are usually located in the left hemisphere of the human brain, is something of a mystery. According to one theory, it's because the left hemisphere is specialized to execute complex sequences of movements, of which speech is an example. In evolutionary terms, brain asymmetries may be advantageous because they enable different tasks to be performed by each hemisphere in parallel, or simultaneously.

The condensed idea
The left and right hemispheres perform different functions

11 Mirror neurons

Mirror neurons are cells that fire during both the execution and observation of a specific action. They have been linked to many behaviours and abilities, from empathy to learning by imitation, as well as implicated in conditions such as autism. Mirror neurons were discovered in monkeys, but it's still not clear whether they also exist in the human brain.

Mirror neurons were identified in the brains of macaque monkeys by a team of Italian researchers during experiments performed in the 1990s. The researchers, who were studying how the brain controls hand and mouth movements, implanted microelectrodes into the monkeys' brains in order to monitor the activity of single cells while the animals reached for pieces of food and put them into their mouths. These experiments revealed that the activity of certain cells increased when the monkeys performed this action.

The cells in question are located in the premotor cortex, a part of the brain involved in planning and executing movements, so the finding was not in itself particularly surprising. By chance, however, the researchers discovered that a few of the same cells also fired weakly when the animals merely observed the researchers putting food into their own mouths, and fired more strongly when they saw other monkeys performing the same action. Subsequently, the same team of researchers identified mirror neurons in several other regions of the monkey brain. They also located cells that fire when monkeys observe an action as well as when they hear the sound related to it.

TIMELINE

1996	1999	2002
Giacomo Rizzolatti and colleagues discover mirror neurons in monkey brains	Vilayanur S. Ramachandran and others propose the broken-mirror hypothesis to explain autism	Rizzolatti and colleagues identify audiovisual mirror neurons in monkeys

But what does it all mean? The precise role of the mirror neuron system in monkeys is still not known, though the researchers who discovered them believe that they perform two functions. First, that mirror neurons are involved in understanding the actions of others – observing an action triggers the mirror neuron system to generate a motor representation of it. This corresponds to the activity produced by the action itself: in other words, the mirror neuron system transforms the visual information into knowledge of the intention of the other's actions. The second proposed function is imitation – or learning to perform an action by observing others.

CASTING A LONG REFLECTION

The discovery of mirror neurons was greeted with a great deal of excitement, and some have hailed it as one of the most important discoveries of modern neuroscience. Since these neurons were discovered in monkeys, researchers have speculated that the human brain may also contain mirror neurons.

In human beings, as in monkeys, mirror neurons are hypothesized to play an important role in imitation and understanding the actions of others. Some researchers argue that they are critical for many aspects of social interactions. These include understanding the intentions of others, and inferring their mental state from their behaviour (an ability referred to as theory of mind); empathy, or putting oneself 'into another's shoes'; self-awareness; and the evolution of, and the ability to learn, language.

Given their purported role in social cognition, one prominent neuroscientist has proposed that a defective mirror neuron system is what causes autism, a neurodevelopmental condition characterized primarily by impairments in social interaction and communication (*see box*). The same researcher argues that the discovery of mirror neurons is 'the single most important "unreported" story of the decade', and has even

2008	**2009**	**2010**
Richard Mooney and team discover mirror neurons in the songbird brain	Alfonso Caramazza fails to find evidence of mirror neuron adaptation	Marco Iacoboni and Itzhak Fried provide first direct evidence of mirror neurons in the human brain

The broken-mirror hypothesis

In the late 1990s, two groups of researchers independently proposed the so-called broken-mirror hypothesis, which states the social impairments characteristic of autism are caused by defects in the mirror neuron system. The broken-mirror hypothesis has received considerable attention in the mass media, but has been the subject of severe criticism by many autism researchers. It is based on assumptions that mirror neurons are involved in understanding action, imitation and language acquisition, and that people with autism are insensitive to the emotions and intentions of others. Critics say both that the first assumption is actually false, and also that there is evidence that people with autism are in fact overly sensitive to others' emotions and intentions. What's more, the broken-mirror hypothesis does not attempt to explain how the mirror neuron system is defective, or how the defects might arise.

referred to the cells as 'the neurons that shaped civilization', because human culture involves the transfer of complex skills and knowledge from person to person.

BUT DO *WE* HAVE MIRROR NEURONS?

Mirror neurons have proven to be highly controversial. A handful of brain-scanning studies show that several regions of the brain are activated during both action execution and observation, and it has been suggested that these areas constitute the human mirror system. But while hundreds of other studies attempt to explain their results by alluding to mirror neurons, very few actually provide hard evidence.

So there is, as yet, very little convincing direct evidence that mirror neurons exist in the human brain. In fact, a number of studies have failed to find evidence of human mirror neurons altogether. In 2009, for example, Harvard researchers exploited a phenomenon called adaptation, whereby neurons reduce their activity in response to the same repetitive stimulus. The researchers showed their participants a film clip of hand gestures and asked them to mimic the action. The scans showed that the cells adapted when the gestures were observed and mimicked, but not when they were mimicked first and then observed.

One of the difficulties is that researchers rarely get the opportunity to examine the working human brain directly. In 2010, though, one research group who had just such an opportunity, while evaluating the brains of conscious epileptic patients about to undergo neurosurgery, claimed that they had obtained the first direct evidence of human mirror neurons. Some of these cells fired both when the patients performed and observed an action, but the activity of almost as many cells decreased during execution and observation, raising doubts that they are indeed mirror neurons. Furthermore, the cells were located in the hippocampus, an area involved in memory formation, and not previously thought to be part of the presumed mirror neuron system.

> **I PREDICT THAT MIRROR NEURONS WILL DO FOR PSYCHOLOGY WHAT DNA DID FOR BIOLOGY.**
> Vilayanur S. Ramachandran, 2000

The researchers who originally discovered mirror neurons in the monkey brain have recently refined their claims, and now suggest that the cells have a far more restricted role than was originally thought. Instead of being involved in understanding the actions of others, they suggest that the cells play a role in helping us to understand, 'from the inside', actions that we already know how to perform. Critics argue that this confirms the alternative theory that mirror neurons are involved instead in selecting and controlling actions.

The condensed idea
Brain cells that 'reflect' the actions of others

12 The connectome

The brain consists of many billions of cells organized into elaborate networks, and neuroscientists are now creating the human connectome – a comprehensive map of all its synaptic connections. Such a map will undoubtedly be useful, but it will not tell us everything we want to know about how the brain works.

The brain contains about 90 billion neurons and is roughly the size of two clenched fists, but it's not the overall size, or the total number of cells, that's important. What matters is how the neurons are connected to each other. The brain consists of a large number of specialized areas, connected by large bundles of nerve fibres called white matter tracts, and neurons within each region form very precise synaptic connections with each other.

It's becoming increasingly clear that a better understanding of the brain will require a detailed knowledge of the connections within and between different parts of the brain – and that the function of an individual neuron is largely determined by the connections it forms with other cells. It is now widely believed, for example, that memory formation involves the strengthening of synapses within distributed networks of neurons, and that conditions such as strokes can be debilitating not only because they kill cells, but also because they disrupt long-distance neuronal connections.

TIMELINE

1931	1985	1986	2005
Max Knoll and Ernst Ruska invent the electron microscope	First use of diffusion tensor imaging (DTI)	Publication of the nematode worm connectome	Olaf Sporns and Patrick Hagmann coin the term 'connectome'

In 2009, the US National Institutes of Health launched the Human Connectome Project. This multimillion-dollar, five-year initiative is analogous to the Human Genome Project, and aims to construct a comprehensive map of the connections in the healthy human brain. This is no mean feat because the brain contains something like one quadrillion synapses. Nevertheless, the hundreds of researchers involved have at their disposal a variety of techniques to help them map the brains of different species at multiple scales.

> **YOU ARE YOUR CONNECTOME.**
> Neuroscientist
> Sebastian Seung, 2011

Eventually, when enough data have been collected, they could be combined to produce a map containing information from the cellular scale all the way up to large-scale connections. Researchers believe that such a map will not only contribute significantly to our understanding of the brain, but will facilitate research into neurological conditions such as Alzheimer's disease, autism and schizophrenia.

TOOLS OF THE BRAIN TRADE

Researchers in this area employ many different methods to study neuronal connectivity.

Electron microscopy: The electron microscope illuminates specially prepared samples of brain tissue with beams of electrons, and can magnify the sample up to ten million times to reveal its most intricate details. In the 1950s, researchers used the technique to produce the first images of synapses, and in the 1980s, it was used to produce the first complete connectome – that of the nematode worm, *Caenorhabditis elegans* (*see page 50*).

Fluorescent dye labelling: Crystals of fluorescent dyes can be applied to slices of brain tissue. These dyes are lipophilic (or 'fat-loving'), and attach themselves to the fatty cell membranes of nerve cells, then travel along the

2007	2008	2009
Jeff Lichtman and Josh Sanes develop the Brainbow method	Ed Callaway and colleagues use a modified rabies virus to trace synaptic connections	Launch of the Human Connectome Project

The worm unturned

Caenorhabditis elegans is a roundworm measuring just 1mm in length and with a nervous system containing a grand total of 304 neurons. Researchers began mapping its connectome in the 1970s, slicing the worms into hundreds of pieces, using electron microscopy to visualize each slice, then reconstructing the pieces to generate a map of all 5,000 connections in the nervous system. The process took more than ten years, and the connectome was finally published in 1986. Many people thought that mapping the *C. elegans* connectome would lead to a huge leap in our understanding of this tiny creature's nervous system, but it did not. Today, nearly 30 years after the connectome was published, we still do not know how the *C. elegans* nervous system – which is very simple compared to our own – generates the worm's behaviours.

nerve fibres when left for a few days. The tissue can then be examined under the microscope to reveal the trajectory of the nerve fibres within it. This technique is often used to examine neural connectivity in animal species, but it can also be used on tissue samples from postmortem human brains.

Diffusion tensor imaging: Diffusion tensor imaging (DTI) is a form of MRI (magnetic resonance imaging) that detects the movements of water molecules along nerve fibres in the living human brain. It can be used to visualize the white matter tracts that form long-distance connections between brain regions.

Genetic methods: Two recently developed methods allow researchers to visualize neuronal networks and synaptic connections in unprecedented detail. Researchers in California used genetic engineering to create modified rabies viruses that produce a fluorescent protein. When inserted into a neuron, it jumps across the synapses to 'highlight' all the other cells with which it forms connections.

The other genetic method, called Brainbow, was developed by researchers at Harvard University, and involves creating genetically engineered mice whose cells contain a set of genes encoding five or six fluorescent proteins of different colours. The genes are organized in such a way that each neuron synthesizes a random combination of two or three fluorescent proteins, thus producing a palette of around 100 different colours. The brain can then be dissected and examined under the microscope, to reveal the fluorescently labelled neurons.

THE GREAT UNKNOWNS

Although methods for detailed analysis of human brain connectivity are not yet available, technological advances will make a comprehensive map possible at some point in the future. Meanwhile, we can learn much from mice and monkeys, whose brains have an anatomical organization similar to that of our own. There's no doubt that a complete human connectome will provide valuable insights into brain function but, as researchers already know from the nematode worm connectome, there are limitations to what it can tell us.

The concept of the connectome ignores the role of genes, and how their interactions with the environment influence our behaviour. It also ignores the role of glial cells, particularly astrocytes, which play important roles in brain function. Another important limitation: although all human brains have the same basic structure, there are significant structural variations between individuals. Consequently, there may be no such thing as a 'typical' brain. Researchers will therefore have to produce the connectomes of as many brains as possible to produce a generalized map of the human brain. Furthermore, the brain is highly dynamic, and its connections are constantly changing. A static connectivity map would not take these changes into account, and it may be necessary to produce numerous connectomes for each individual, to determine exactly how neural connections change with time.

The condensed idea
A complete wiring
diagram of the brain

13 Embodied cognition

Traditionally, the brain is thought of as the master controller – generating thoughts and actions by converting abstract representations of the world into commands for the body. According to a new theory, however, thoughts and behaviours are not produced by the brain alone, but are the result of dynamic interactions between the brain, the body and the environment.

The brain is usually viewed as an organ that controls behaviour by processing perceptions of the outside world and transforming them into mental representations which are then used to guide thoughts and behaviour. In other words, the brain acts to process abstract information, and our knowledge of the world resides in memory systems that are separate from action and perception.

The embodied cognition hypothesis is a radical new idea that rejects this traditional view. It regards the brain as one component of a wider system that also includes the body and the environment, both of which are critical for shaping our thoughts, emotions and behaviours. Thus, our mental representations are 'embodied', or grounded, in the physical state of the body and its interactions with the environment. As such, they are intimately linked to the brain's sensory and motor systems.

The idea of embodied cognition originates in continental philosophy. Immanuel Kant believed that the mind was distinct from the body, but that the

TIMELINE

1755	1927	1980	1998
Immanuel Kant writes *Universal Natural History*	Martin Heidegger writes *Being and Time*	Publication of George Lakoff and Mark Johnson's book *Metaphors We Live By*	Andy Clark and David Chalmers propose the extended mind hypothesis

two were closely related; he also believed that our ability to think is dependent upon the properties of our bodies. For Kant, bodily movements are necessary for thinking, and for recalling and connecting mental representations. Nearly two centuries later, Martin Heidegger argued that we experience the world by interacting with it, and that thinking involves putting things to use. Likewise, Maurice Merleau-Ponty believed that the body is not just an object of perception, but is also crucial for it.

METAPHORICAL GROUNDING

Early embodied cognition theorists emphasized the effects of the body on thought processes. They argued that language is closely linked to the perception of the body and that the metaphors we use are shaped to the form of our bodies. Thus, our experiences are grounded in metaphorical thoughts,

Mind-expanding stuff

The extended mind hypothesis is a radical form of the embodied cognition hypothesis, a modern philosophical concept that simply asks, 'Where does the mind stop and the rest of the world begin?' As its name suggests, the extended mind hypothesis states that the mind, and the thought processes related to it, are not confined to the brain but also extend to certain elements of the outside world.

Proponents of the extended mind theory argue that human beings are closely coupled to external components to create a larger thinking system. Computers and the Internet are prime examples of external components of the mind that we use, among other things, to reduce the memory load placed on the brain. Recent research shows that people are less likely to commit information to memory, and therefore to recall it later, when they know it's available online. In other words, the Web is like an external hard drive that we can use to store information.

2005

Jessica Witt and colleagues begin studies on how sporting performance affects perception

2006

Thorsten Hansen and team show that memories of an object's colour influence colour perception

2011

Daniel Wegner and colleagues publish a study examining Google's effects on memory which are in turn based on how we use our bodies to interact with the world. Similarly, we often express emotions in terms of movements or positions in space. Positive emotions are always associated with upward movements (we may say that our 'spirit soared', or that we are feeling 'on top of the world'), while negative thoughts are linked with downward movements (for example, feeling 'down in the dumps').

Numerous studies support this view, showing that bodily states can indeed strongly influence, or directly cause, mental states. For example, you perceive people as being friendlier when you are holding a warm cup of coffee than when you are holding a cold one; you are more likely to wash your hands after thinking about bad deeds than about good ones; and you will perceive a heavy book as being more important than a lighter one. This type of research shows the importance of embodied metaphorical thought – affection is warmth; immoral actions are dirty whereas moral acts are pure; and important subjects are 'weighty'.

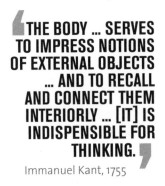

THE BODY ... SERVES TO IMPRESS NOTIONS OF EXTERNAL OBJECTS ... AND TO RECALL AND CONNECT THEM INTERIORLY ... [IT] IS INDISPENSIBLE FOR THINKING.

Immanuel Kant, 1755

According to some researchers, however, these studies are just the tip of the iceberg, and do not demonstrate 'true' embodied cognition. They argue that behaviour emerges from dynamic interactions not only between the brain and the body, but also the external environment. This view emphasizes the importance of our actions and perceptions, as well as the process of simulation, or re-enactment of actions and perceptions. Far from being abstract, the brain's mental representations are thus closely tied to – and dependent upon – the events that we experience.

ACTION, PERCEPTION, SIMULATION

Various psychological studies show that our actions and their consequences feed back to influence our perceptions, and that the process of simulation plays a key role in this. When we perceive an object, it is stored in our visual memory. Encountering similar objects later on triggers these memories to simulate the earlier perceptions, and these simulations can interfere with our present perceptions. For example, seeing a colour picture of a banana closely followed by a grey-scale version alters our perception towards the natural colour of the object, so that the banana in the second image appears bluish.

> **BODILY STATES AND ACTION UNDERLIE COGNITION.**
>
> American cognitive scientist
> Lawrence Barsalou, 2008

Action and perception are closely coordinated, and the two are linked by simulation. When we see an object, the brain prepares to use it by running the appropriate simulations of possible actions. Seeing a cup activates a simulation of grasping; seeing a hammer activates a power grip; and seeing a grape activates a precision grip. These simulations are dependent on the position of the object with respect to the body – seeing the object in the right orientation facilitates action simulation, whereas seeing it the wrong way round hinders it, and makes the object seem farther away than it actually is. Simulation also interferes with the ability to perform unrelated movements in the split second following perception of the object.

The consequences of our actions influence our perceptions, too. During a winning streak, athletes often say that they perceive their targets as being bigger than they actually are, and this is backed up by research. Golfers and high-school American football players perceive the size of the hole or goal as bigger when they are performing well, making it seem easier to score. The reverse is also true – when performing badly, they see the goal as smaller, making it seem more difficult. The way we perceive a task is also influenced by the amount of effort required to perform it. An uphill location seems farther away than it actually is, and a hill is perceived as being steeper than it really is if we are carrying something heavy.

The condensed idea
The mind is a product of brain, body and outside world interactions

14 Bodily awareness

Who am I, and how did I come to be who I am? Self-identity is a complex phenomenon that consists of multiple components, including personality, memory and sexual and national identity. At the core of the self is something that most of us take for granted – the body, and our awareness of it.

Existentialist philosophers such as Friedrich Nietzsche and Maurice Merleau-Ponty knew that what we call the 'self' is intimately connected to the body, and neuroscience is finally beginning to catch up. Modern research has begun to reveal the neurological basis of bodily recognition, and to show that such awareness of one's body is a core component of self-identity.

Investigations of bodily awareness began over a hundred years ago, with the work of the British neurologists Henry Head and Gordon Holmes. They studied dozens of patients in whom bodily awareness was disturbed following damage to the right parietal lobe, and concluded that this part of the brain contains a dynamic representation of the body, or a model of the self, which they referred to as the 'body schema'.

Modern research has revealed that the brain does indeed encode multiple representations of the body, and that bodily awareness consists of two related but distinct components. Each of these components can be distorted or manipulated separately, but both are critically important for self-consciousness.

TIMELINE

1910	1946	1998
Head and Holmes postulate the concept of the body schema	Publication of Merleau-Ponty's *The Phenomenology of Perception*	Matthew Botvinick and Jonathan Cohen describe the rubber hand illusion

BODY OWNERSHIP

We usually experience ourselves as being located within our bodies, and recognize our bodies as our own. This is because the brain distinguishes between what is part of one's self and what is not, giving rise to the sense of body ownership, one of the components of bodily awareness. But sometimes body ownership is disrupted – for example, during out-of-body experiences (in which the 'self' appears to leave the physical body temporarily), in a delusion called somatoparaphrenia (in which patients deny that a limb belongs to them) and in a condition called body integrity identity disorder (in which individuals have a burning desire to amputate a healthy limb).

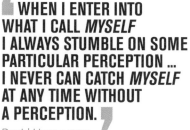

WHEN I ENTER INTO WHAT I CALL *MYSELF* I ALWAYS STUMBLE ON SOME PARTICULAR PERCEPTION ... I NEVER CAN CATCH *MYSELF* AT ANY TIME WITHOUT A PERCEPTION.

David Hume, 1739

Distortions in the sense of body ownership are not only limited to psychiatric conditions, however, but also occur in everyday life. Head and Holmes suggested that 'anything which participates in the conscious movement of our bodies is added to the model of ourselves', and that a woman's body schema 'may extend to the feather in her hat'. We now know that using a computer mouse or a tool causes the object to be incorporated into the brain's representation of the body. In other words, the brain comes to regard foreign objects as a part of the 'self' after prolonged use. This explains why a blind man can 'feel' with his cane. Crucially, it will eventually lead to a new generation of artificial limbs that the brain recognizes to be part of the body.

AGENCY AND FREE WILL

The other component of bodily awareness is agency, that is, the sense that we are in control of our own bodies and are responsible for our actions. Agency depends upon so-called forward models generated in the brain, which predict the consequences of any actions. When you decide to turn on a light, for example, your brain predicts that you will reach for and then touch the

Can you trust your eyes?

Simple manipulations of the sensory inputs to the brain give rise to illusions that alter its representation of the body and distort the sense of self. In the rubber hand illusion, touch sensations are perceived to arise from an artificial hand. The illusion is induced by simultaneously touching the rubber hand and the real one in exactly the same way, while the participant looks at the rubber hand. This creates a discrepancy of sensations that tricks the brain into thinking that the rubber hand is part of the body, and shows that vision is the most important sense for body ownership.

The body swap illusion is based on the same principles. Here two people standing opposite each other, both wearing head-mounted visual displays with cameras attached to them. The cameras feed into the other person's visual display, so that both people see their own body from a third-person perspective, located in the position of the other person's body. When both people are touched simultaneously, they perceive the sensations as originating from the other's body.

switch on the wall, and that you will then hear a click and see the room light up. These predictions are then compared to what actually happens, and a close correspondence between the two gives us the sense that we are in control of our actions.

Research into the underlying mechanisms suggests that the brain actively distorts time perception to give rise to the sense of agency, to make us think that we are in control. Some experiments suggest that voluntary movements are perceived as occurring later than they do, whereas their effects are perceived as occurring earlier than they actually do, so that the intention of an action and its effects are perceived as occurring together.

Other studies have found that the brain reverses the order of events, so that the effects are perceived a fraction of a second before the action that caused them. There is a delay of about 80 milliseconds between when something happens and when we perceive it, because of the time taken by the brain to process information. According to these results, the brain recalibrates the timing of events to make them consistent with our expectations.

The sense of agency is important because it leads to the conscious experience of 'free will', and it also contributes to the sense of body ownership by confirming that the hand performing the action belongs to you. In some conditions, however, the sense of agency is disrupted: in alien hand syndrome, a rare psychiatric disorder, patients deny that they are in control of the actions of one of their upper limbs; and in schizophrenia, patients often misattribute their thoughts and actions to external, controlling forces.

SEEING OURSELVES

Bodily awareness arises through something called multisensory integration – in real terms, the brain combines three different types of sensory inputs that it receives from the body: visual information, touch information and proprioceptive information (relating to the position of the limbs in space). Visual information is processed in the occipital lobe, whereas touch and proprioception inputs are processed in the somatosensory cortex. The three are then combined in the superior parietal lobule, to generate a dynamic representation of the body that is variously referred to as the *body schema* or *body image*.

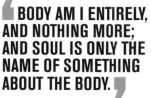

BODY AM I ENTIRELY, AND NOTHING MORE; AND SOUL IS ONLY THE NAME OF SOMETHING ABOUT THE BODY.
Friedrich Nietzsche, 1883

This representation manifests itself as a mental picture of the body, and it is through this picture that we perceive our bodies. Disturbances in awareness of the body can change the brain's representation of it, which in turn can profoundly alter the sense of self. For example, discrepancies between body image and the physical form of the body can produce cognitive dissonance (conflicting thoughts and feelings that are psychologically distressing), and may contribute to, for instance, anorexia, body dysmorphic disorder or transsexualism.

The condensed idea
Awareness of the body is critical to the sense of self

15 Free will

Free will is central to our idea of what it means to be human, and is a subject that has puzzled philosophers for centuries. We all want to believe that we have free will, that we are in control of our actions and decisions, but brain research suggests that this may be nothing more than an illusion.

One age-old debate about human behaviour divided philosophers into two camps. Advocates of free will, such as René Descartes, believed that we are rational agents who can choose our actions, whereas determinists, such as John Locke, argued that our choices are constrained by the physical forces governing our bodies. About 30 years ago, neuroscientists joined the debate, following the publication of a study suggesting that the choices we make are the result of unconscious brain processes. Although this has been interpreted to mean that we do not have free will, not all neuroscientists see it that way.

FREE WON'T?

Early evidence against free will comes from a classic 1983 study that measured the brain activity associated with voluntary hand movements. In a relatively simple experiment, researchers asked study participants to move their fingers whenever they felt the urge to do so. They also had to look at a blank clock face with a dot moving around it, and note the position of the dot when they became aware of the intention to move.

The researchers used electroencephalography (EEG) to record participants' brain activity, and detected a signal called the 'readiness potential' in the supplementary motor area, a part of the frontal cortex involved in planning

TIMELINE

1689	1964	1983
John Locke argues against free will	Hans Kornhuber and Lüder Deecke identify the readiness potential	Benjamin Libet et al. report that the readiness potential precedes awareness of the intention to move

movements. Surprisingly, they detected this signal about a third of a second before the participants reported becoming aware of the intention to move their fingers.

Several other research groups have obtained similar results using modern techniques. In 2008, researchers in London used functional magnetic resonance imaging (fMRI) to scan participants' brains while they decided between pressing two buttons, using either the left or the right index finger. A series of letters appeared on a small screen inside the scanner, and the participants had to note which was displayed at the time they made the decision to press one of the buttons. The researchers found that motor cortex activity could predict, with an accuracy of around 60 per cent, which of the two buttons the participants would press, up to 10 seconds before they became aware of the intention to act.

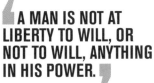

A MAN IS NOT AT LIBERTY TO WILL, OR NOT TO WILL, ANYTHING IN HIS POWER.

John Locke, 1689

More recently, the original findings were confirmed yet again by a team of American neurosurgeons, who used electrodes to record directly from neurons in the brains of epilepsy patients while they performed self-initiated finger movements. They found that single cells in the motor cortex become active up to a second and a half before the patients reported making the decision to move.

Some of the research in this area shows that activity in the frontal cortex is stronger when people prepare to carry out an action and then intentionally stop themselves from doing it, than when they prepare and perform the same action. On the basis of this 'veto power', some have suggested that a better term for free will would be 'free won't'.

ARE YOU READY?
The original findings proved to be highly controversial, and have been debated at great length ever since they were published. The results of the study, and

2007	2008	2010	2011
Patrick Haggard et al. discover the brain's veto power	Researchers predict choices from brain activity up to 10 seconds before actions	Study shows readiness potential can be detected regardless of which decision is made	Neurosurgeons identify motor cortical neurons that are activated up to 1.5 seconds before the will to act

of those that followed, seem to show that the brain prepares our actions before we make the conscious decision to act. In other words, our actions and decisions are determined by brain mechanisms of which we are not aware. This directly contradicts the classical notion of free will, according to which we are free to decide between different courses of action.

The research has been widely criticized, however. The main problem is that the studies rely on participants' perception of time and their subjective reports of the timing of events. These events occur within fractions of a second, making it extremely difficult to pinpoint exactly when they occurred. This is further complicated by the brain's processing time – it takes a fraction of a second to interpret visual information, and another fraction to produce a motor output. The time periods are so small that in everyday life these processes seem to happen instantaneously, but in experiments such as these, they can make a huge difference.

Another problem: it's not entirely clear exactly what the readiness potential is. Until recently, it was widely thought to be a neural signature of the planning, preparation and initiation of voluntary movements, a gradual

Alien hand syndrome

In Stanley Kubrick's classic 1964 black comedy *Dr Strangelove*, the lead character, played by Peter Sellers, has a right hand that acts of its own volition. At times, it clutches his throat and he has to use his other hand to bring it under control, and in the famous final scene, it performs a Nazi salute. This is a fictional depiction of alien hand syndrome, a neuropsychiatric disorder in which a patient's hand seemingly takes on a mind of its own and is not under voluntary control.

Alien hand syndrome has been documented in split-brain patients (*see page 41*), who have had the connections between the left and right hemisphere severed to control epilepsy, and can occur following a stroke or infection. It is also associated with damage to the supplementary motor cortex. In real life, patients often view the affect limb as being "disobedient," and sometimes even believe that is possessed by an external force.

build-up of neuronal activity that occurs in the premotor cortex in the seconds prior to executing an action. But several studies suggest otherwise. In one of these, researchers from New Zealand used EEG to compare the brain activity preceding the decision to move with activity preceding the decision not to move. They detected the same pattern of electrical activity in both cases, suggesting that the readiness potential does not represent the brain preparing to move.

Another study, published in 2012, provides an alternative explanation for the readiness potential. It's known that decisions based on visual information involve the accumulation of inputs within separate neuronal networks, and that the decision is based on the network with the strongest activity. French researchers hypothesized that something similar might happen during decisions to make voluntary movements. They therefore repeated the original experiment, but also asked their participants to act immediately if they heard a click while waiting. Using EEG to measure brain activity during the task, they found that those with the fastest responses to the clicks also exhibited the largest readiness potentials. This suggests that the readiness potential and the decision to move are related to neural noise, or spontaneous fluctuations in neuronal activity.

WE FEEL WE CHOOSE, BUT WE DON'T.
Patrick Haggard, 2011

So although some brain research seems to suggest that we do not have free will, this is by no means conclusive, as the findings are still open to interpretation.

The condensed idea
Free will may be an illusion

16 Sex differences

Subtle observable differences exist between male and female brains, but how exactly these relate to differences in behaviour is unknown. Such gender variations in the brain are often exaggerated and misappropriated, not only by the mass media but also by scientists, to reinforce stereotypes and perpetuate myths.

The science of sex differences has always been – and still is – fraught with controversy. Some believe that behavioural differences between men and women are mostly due to cultural influences, while others argue that sex differences are largely determined by biology. In reality, the situation is far more complex. It lies somewhere in the middle, and involves two related but independent factors, which are often confused or conflated.

One of these factors is biological sex, which is determined by chromosomes. Most people have either two X chromosomes, which makes them female, or one X and one Y chromosome, which makes them male. The other is gender, which is influenced largely by the socialization process. As we grow up, we learn society's norms about how males and females look and act; for most people, sex and gender are matched, and so they inadvertently conform to these norms.

Men and women's brains differ in subtle ways, and these differences are probably established in the womb, due to the effects of sex hormones, which masculinize or feminize the organ as it develops. However, we still do not understand the effects of sex hormones on the developing brain, or how the subtle differences observed between men and women's brains are related to differences in their behaviour.

TIMELINE

1992	1997
Simon LeVay reports differences in the hypothalamus of heterosexual and homosexual men	Simon Baron-Cohen proposes the extreme male brain hypothesis

BATTLE OF THE SEXES?

The most obvious difference between the brains of men and women is overall size – men's brains are, on average, between 10 and 15 per cent larger than women's. In one recent study, neuroscientists compared the brains of 42 men and 58 women postmortem, and found that men's weighed an average of 1,378g (3lb), compared with 1,248g (2¾lb) for women. These size differences have been found repeatedly, but they emerge only when comparing large numbers of people, so some women's brains are larger than the average whereas some men's are smaller. These differences partly reflect the fact that men are generally bigger and taller than women, but they are not related to differences in intelligence.

Men and women's brains also differ in overall composition. Male brains tend to have a slightly higher proportion of white matter, whereas those of females have a higher proportion of grey matter in most parts of the cerebral cortex. Consequently, the cortex is slightly thicker in women's brains than in men's and, according to several studies, is slightly more convoluted as well. There are also sex differences in the size of individual brain structures. The hippocampus, a structure involved in memory formation, is on average larger in men than in women, as is the amygdala, which is also involved in memory, as well as emotions.

Another sexual variation is found in a structure called the third interstitial nucleus of the anterior hypothalamus. The function of this tiny structure

Factoring in disease

Women are twice as likely to develop depression than men. This is probably because of a mixture of biological factors, such as the differential effects of sex hormones on the brain – and cultural elements, such as sexual inequalities and the pressures of juggling work and family. A recent study also shows that women's mental abilities deteriorate more rapidly with Alzheimer's disease than men's.

2000
Researchers report that women are better at attending to sounds played into both ears at once

2008
Dick Swaab and colleagues report 'sex reversal' in the hypothalamus of transsexuals

2012
Researchers report that women with Alzheimer's disease deteriorate more rapidly than men

The extreme male brain hypothesis

People with autism tend to perform poorly on tests of empathizing, or the ability to put oneself in somebody else's shoes, but do well on tests of systematizing, or finding repeating patterns. Similarly, women tend to score higher on the empathy scale, and men on the systematizing scale. These observations led one researcher to propose the highly controversial 'extreme male brain' hypothesis of autism. The hypothesis states that autism is an extreme form of the normal male cognitive profile, which occurs as a result of high testosterone levels in the womb. Accordingly, people with autism can be considered as 'hyper-systematizers' who focus more on patterns and fine details than on other people's thoughts and actions. The extreme male brain hypothesis has been used as an explanation for why autism is four times more prevalent in males than in females, and why people with autism can excel in disciplines such as maths and engineering.

is unknown, but research from four different laboratories has repeatedly found that it is almost twice as large in males than in females. It has also been linked to sexual orientation and gender identity: one study showed that it is more than twice as large in heterosexual males than in homosexual males, where it more closely resembles that of women; another found that it is smaller in male-to-female transsexuals, and larger in female-to-male transsexuals. These studies have been criticized for their small sample sizes, and the findings have not been confirmed.

STEREOTYPES AND MYTHS

Numerous studies show subtle differences in male and female behaviour and in cognitive functions, too. Men tend to be more aggressive and outperform women on mental tasks involving spatial skills such as mental rotation, whereas women tend to be more empathetic and perform better on verbal memory and language tasks. Findings like these are often exaggerated to reinforce the stereotypes that women are bad at reverse parking and that they love to chat!

In some cases, individual studies purporting to show sex differences in certain tasks are misappropriated. For example, according to a tiny postmortem study published in 1982, the corpus callosum, the massive bundle of nerve fibres connecting the two brain hemispheres, is proportionally larger in women than in men. This was widely reported to mean that women are better at multitasking, even though subsequent work has failed to replicate the results. A more recent study showed that women are marginally better than men at paying attention to sounds presented to both ears simultaneously – this was interpreted by some as evidence that 'men don't listen'.

Many of these claims are accompanied by the assertion that the observed differences between men and women's brains are 'hard-wired' and, therefore, irreversible. We now know, however, that brain structure and function change in response to experience, so any observed differences between the brains of men and women could also be due to differences in upbringing and socialization. To date, though, very little research has been done to investigate how different nurturing styles might influence brain development.

> **THE SHEER COMPLEXITY OF THE BRAIN MAKES INTERPRETING AND UNDERSTANDING THE MEANING OF ANY SEX DIFFERENCES WE FIND A VERY DIFFICULT TASK.**
> Australian psychologist Cordelia Fine, 2010

The condensed idea
We don't know how male/female brain differences affect behaviour

17 Personality

Personality refers to a set of mental, emotional and social characteristics that are assumed to remain stable over time. These characteristics, more than anything else, determine how people define themselves. Traditionally, the study of personality was carried out by psychologists. In recent years, however, neuroscientists have begun to investigate how personality traits might be linked to brain structure and function.

The study of personality aims to determine why people are the way they are, and how they differ from one another. In the past, this was left to psychologists, who were mainly concerned with describing personality rather than explaining it, using various tests to measure how people scored on different personality traits. It's generally agreed, however, that all psychological processes have a basis in the brain. Personality must, therefore, be determined by brain structure and function, but we still have no idea how. Nevertheless, some researchers have now begun to investigate the biological basis of personality.

THE 'BIG FIVE' TRAITS

Over the years there have been numerous attempts to classify personality types on the basis of various traits. According to one influential theory of personality, developed by Hans Eysenck in the mid-20th century, personality consists of two main dimensions – extroversion and neuroticism. Subsequently, Eysenck added psychoticism as a third dimension. Psychologists now generally agree that there are five main personality traits:

TIMELINE

1921	1947	1985
Carl Jung popularizes the terms 'extroversion' and 'introversion'	Publication of Hans Eysenck's book *Dimensions of Personality*	Paul Costa and Robert McCrae publish a test to measure the Big Five personality traits

Extroversion is defined as the 'act of being predominantly concerned with and obtaining gratification from what is outside the self', and exists on a continuum with introversion, or a 'tendency towards being wholly or predominantly concerned with and interested in one's own mental life'. Extroverts tend to be outgoing and gregarious, taking pleasure in social events, whereas introverts tend to be quiet and reserved, and prefer solitary activities such as reading.

Neuroticism manifests itself as anxiety, moodiness and irritability. People who score highly on neuroticism are often shy and self-conscious. They are highly sensitive to negative emotions, and can therefore perceive otherwise normal situations as threatening. They also have a higher risk of developing depression.

Conscientiousness is a trait that equates roughly to will power, and is manifested in behaviours such as efficiency, neatness and deliberation. Conscientious individuals tend to be reliable and hard working, whereas those who score low on this trait tend to be easy-going and less goal-driven.

Agreeableness is defined by behaviours that are thought of as being kind, considerate, cooperative and sympathetic. People who score highly on this trait tend to be altruistic, modest and trustworthy, while those who score low tend to be less concerned with others' well-being and have less empathy.

Openness is a trait characterized primarily by creativity, imagination and intellectual curiosity. People who score highly on openness tend to actively seek novel experiences and have liberal political views, whereas those who score low tend to be closed to new experiences and have conservative political views.

1999	2011	2012
Debra Johnson and colleagues find differences in cerebral blood flow between extroverts and introverts	Colin DeYoung and team publish correlations between personality traits and functional connectivity	Mehmet Mahmut and Richard Stevenson link psychopathy to decreased smell sensitivity

BELOW THE SURFACE

Psychologists typically study personality by asking subjects to fill out questionnaires designed to measure how they score on the various personality traits. Neuroscientists are now beginning to use various methods to understand the biological basis of personality. One method involves looking for genetic variants associated with personality traits. For example, a 2005 study showed that variations in the catechol-O-methyltransferase gene (which encodes an enzyme that regulates dopamine activity) is associated with openness. Another showed that variants of the monoamine oxidase-A gene (which encodes an enzyme that breaks down the neurotransmitters dopamine, serotonin, adrenaline and noradrenaline) is associated with differences in agreeableness and conscientiousness.

Another method uses brain-scanning techniques to look at individual differences in brain structure and function. One early study used positron emission tomography (PET) scanning to show that introverts have greater blood flow to the frontal lobes and the front region of the thalamus than extroverts; researchers suggest that this is related to introverts' propensity for solitude and self-reflection. Conversely, the same study showed that extroverts exhibited increased blood flow to the anterior cingulate gyrus and back end of the thalamus; this, the researchers argue, may be related to their need for high levels of sensory and emotional stimulation. Another study showed that extroverts show increased activity in midbrain dopamine-producing pathways in response to positive images, which suggests that they may be highly sensitive to rewards, particularly those associated with social interactions.

Psychopathy

Psychopathy is a personality disorder that has garnered much interest in recent years. Psychopaths tend to show little emotion, especially fear, and lack empathy. They can be callous, manipulative and impulsive and, although they seem to know right from wrong, show little regard for the consequences of their actions. Psychopathy is associated with a reduction in the volume of the amygdala, a brain structure involved in fear and other emotions, and with abnormal activity in the orbitofrontal cortex, which is thought to be involved in decision-making and in evaluating the rewards and punishments of our actions. People who score high on the psychopathic trait seem to have abnormal white matter tracts connecting these regions, too. The orbitofrontal cortex is also involved in processing smell information, and one recent study showed that people with psychopathic traits have trouble naming smells and distinguishing between them.

Another research group recently reported that the five big personality traits are correlated with differences in connectivity between brain regions. Using diffusion tensor imaging (DTI), they found that neuroticism is associated with increased connectivity between the limbic system and a structure called the precuneus, which may reflect a greater propensity to combine social cues with emotional information. Extroversion was linked to increased connectivity between parts of the limbic system associated with reward and motivation, and with increased connectivity of a brain region important for facial recognition. Openness was correlated with enhanced connections between regions known to be involved in daydreaming and imagination. Agreeableness was linked to increased connectivity between a set of regions collectively thought to be involved in social and emotional attention. And conscientiousness was found to be associated with increased connectivity within the medial temporal lobe, which contains structures involved in recalling the past and imagining the future.

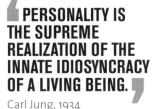

PERSONALITY IS THE SUPREME REALIZATION OF THE INNATE IDIOSYNCRACY OF A LIVING BEING.
Carl Jung, 1934

Most of these studies involve small numbers of participants and find relatively weak correlations between personality traits and certain patterns of activity or connections. They are all based on averaged results from the samples used, so individual brain scans showing similar patterns don't necessarily predict personality. Nevertheless, they are exciting because they represent the beginnings of the neuroscientific study of personality, and future research will undoubtedly lead to a clearer understanding of how personality is linked to brain structure and function.

The condensed idea
How does brain structure and function determine personality?

18 Brain-damaged patients

Neuroscience relies on studies involving large numbers of participants, but another way to investigate how the brain works is to examine the behaviour of brain-damaged individuals. The history of neuroscience is peppered with cases of people who underwent dramatic changes as a result of brain damage. Such case studies are the cornerstones on which modern neuroscience is based.

One of the earliest – and best-known – case studies in neuroscience and psychology is that of a railroad worker named Phineas Gage. One day in September 1848, Gage inadvertently caused an explosion that propelled a metre-long iron rod through his head at high speed. The rod was later recovered about 10m (33ft) from the scene of the accident, 'smeared with blood and brain'.

Remarkably, Gage survived this horrific ordeal and regained consciousness shortly after the accident. He made a speedy recovery, and went back to work, apparently unaffected. In the following months, however, Gage's family, close friends and colleagues began to notice something different about him. Once diligent and polite, Gage became irrational, ill-tempered and foul-mouthed, and began making sexual advances towards any woman he encountered.

The iron bar had destroyed much of Gage's left frontal cortex as it passed through his head. This affected his decision-making abilities, and made him lose all of his social inhibitions. According to popular accounts, the result

TIMELINE

1848	1861	1874
Phineas Gage suffers frontal lobe damage during a work accident	Pierre Paul Broca describes brain damage in stroke patients who cannot speak	Carl Wernicke describes brain damage in stroke patients who cannot understand speech

Speech and strokes

In the 1860s and 1870s, the physicians Pierre Paul Broca and Carl Wernicke worked with stroke patients who had speech deficits, then examined their brains when they died (*see also Chapter 28: Language processing*). Broca's patients were all unable to produce speech, and all had damage to the same region of the left frontal lobe. The patients examined by Wernicke had difficulty understanding spoken language, and had damage to an area in the left temporal lobe. These studies provided scientific evidence for the idea that the left and right hemispheres perform different functions, and also led to the idea of cerebral dominance. The left hemisphere is said to be dominant, because it contains the speech centres. Crucially, this classic model of Broca's and Wernicke's Areas being involved respectively in speech production and comprehension is overly simplistic. Both areas perform other functions, too, and their exact location is still being debated.

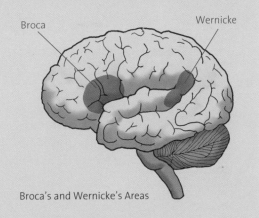

Broca's and Wernicke's Areas

was a profound change in Gage's personality – those who knew him said he seemed like a different person, and that he was 'no longer Gage'. Following his accident, Gage held down several menial jobs, and toured with a circus, in which he was exhibited, together with the iron rod that impaled his head. He died near San Francisco in 1860.

Ultimately, the stories about him cannot be verified. The case of Phineas Gage has become something of a legend, and it is impossible to separate the myth from reality. Researchers have tried to reconstruct Gage's skull in order to determine the extent of the damage, but the accuracy of the reconstructions

1940	1953	1957	1981
William P. van Wagenen performs the first split-brain operations	H.M. undergoes neurosurgery for severe epilepsy	Brenda Milner and William Scoville publish paper on H.M.'s memory function	Roger Sperry wins the Nobel Prize for his work on split-brain patients

is open to question. Furthermore, recent research suggests that Gage's reported behavioural changes may have lasted only a short time after his injury. Nevertheless, it was because of this case that mental abilities such as decision-making and social cognition came to be associated with the frontal lobes.

THE REVEALING CASE OF H.M.

Another well-known case is that of the amnesic patient known as H.M., who had neurosurgery as a treatment for severe, drug-resistant epilepsy. During the operation, surgeons removed the hippocampus from both sides of H.M.'s brain. The operation cured H.M.'s seizures, but had severe consequences – he lost the ability to form new autobiographical memories, and therefore lost an essential component of his self-identity.

At the time, researchers suspected that this structure was involved in memory, but this was not yet clear, and the case of H.M. showed unequivocally that the hippocampus is critical for memory formation. Subsequently, a researcher named Brenda Milner performed exhaustive assessments of his memory and, in the process, single-handedly founded the discipline of neuropsychology.

❝THE STUDY OF H.M. BY BRENDA MILNER STANDS AS ONE OF THE GREAT MILESTONES IN THE HISTORY OF MODERN NEUROSCIENCE.❞
American neuroscientist Eric Kandel, 2008

Milner's studies revealed that there are distinct types of memory. The surgery had left H.M. with severe anterograde amnesia, or an inability to form new memories of life events. But Milner's tests showed he could retain small amounts of information for short periods of time, indicating that memory consists of separate short-term and long-term stores. Subsequent tests by Milner and Suzanne Corkin found that H.M. could learn simple motor skills and draw a detailed map of his house, suggesting that these abilities depend on other distinct memory systems.

H.M. died in 2008, and his full name was revealed as Henry Gustav Molaison. He pledged his brain for research, and when he died, the organ was removed and taken to California, where it was cut into thousands of slices. The data are currently being digitized and placed online for neuroscientists to use. Arguably, H.M. contributed more to our understanding of memory than any other person, and he will undoubtedly continue to teach us more.

SEVERING THE CONNECTION

In the 1940s, neurosurgeons in New York developed a radical procedure to treat patients with severe epilepsy. The method involved cutting the corpus callosum, the enormous bundle of nerve fibres that connects the left and right hemispheres of the brain. This prevents the two sides from communicating with each other and reduces the severity of the epileptic seizures.

The procedure, called a corpus callosotomy, was performed on fewer than a dozen patients, who were thoroughly examined in a series of experiments during the 1960s. Remarkably, these so-called split-brain patients can lead normal lives, albeit with some bizarre quirks in their behaviour. They would, for example, choose a garment from the wardrobe with one hand, then throw it back with the other, as if they were two different people, each with their own thoughts and intentions.

In the laboratory, other consequences of the surgery emerged. If, for example, they see a word such as 'pen' in their left visual field, the information enters the right hemisphere, but cannot be transferred to the left. The patient therefore cannot say the word, because the speech centres are located in the left hemisphere. They can, however, pick out the object with their left hand, which is controlled by the right hemisphere, because the connections between the brain and the hand remain intact.

The studies of split-brain patients added weight to the idea that each brain hemisphere is specialized, but it also contributed to the popular myth that the 'left brain' is logical and the 'right brain' creative. Crucially, split-brain patients show that the left and right hemispheres work together. The brain does contain specialized areas, but it functions as an integrated whole, and the connections between the two hemispheres, and between areas within the same hemisphere, are vitally important.

The condensed idea
Case studies provide valuable insights into brain function

19 The theatre of consciousness

We experience ourselves and the world as a constant flow of thoughts and sensations, but how the brain generates this stream of consciousness is a mystery. According to one influential theory, consciousness is like a theatre – a 'spotlight of attention' shines a bright beam onto certain neural processes, and those that are lit up enter the 'stage' of conscious awareness.

Consciousness is something with which we are all very familiar, but at the same time it is a deeply mysterious phenomenon. We all know what it means to be conscious – right now, you are conscious of the words on this page, but you may also be aware of the sensations arising from the chair you are sitting in and, perhaps, of some background noise. And yet, despite centuries of speculation by philosophers and, more recently, neuroscientists, we still have little idea of what consciousness actually *is*, or how the brain generates it.

The brain is continuously performing multiple operations, but at any given moment we have access to, or are consciously aware of, only a small number of them. We might focus our attention on something going on in the world around us, such as an interesting programme on the television, then switch to an internally generated thought or memory. These are the contents of consciousness, which are experienced as a constant flow of perceptions. One influential theory, the global workspace model, explains how the brain achieves this to produce this 'stream of consciousness'.

TIMELINE

1641	1960	1983
Publication of *Meditations on First Philosophy* by René Descartes	George Miller, Eugene Galanter and Karl Pribram coin the term 'working memory'	Bernard Baars proposes the global workspace model of consciousness

SCREEN SHOTS

The global workspace model uses the theatre as a metaphor for the brain mechanisms responsible for generating consciousness. It regards the brain as a distributed, parallel system containing multiple processors that operate simultaneously. These processors are the actors, all of which can be made available to consciousness. We are, however, conscious of those actors only when they step onto the 'stage'. When they are not on stage, their actions are being performed unconsciously.

In the theatre of consciousness, the stage corresponds to working memory, which allows us to retain and manipulate limited amounts of information for short periods of time. It can be thought of as a kind of screen, onto which experiences are projected to your 'mind's eye'. Selective attention directs the

Fame in the brain

The French philosopher René Descartes believed that the brain generates consciousness by selecting certain information and displaying it on an internal screen, where it is viewed by a homunculus (meaning 'little man', a metaphor for the soul). Neuroscientists reject this idea, but the global workspace and other theories do assume that there is a special place in the brain where consciousness 'happens'. The modern-day philosopher Daniel Dennett therefore refers to them critically as 'Cartesian theatre'

models, and has proposed the 'multiple drafts' model as an alternative. According to Dennett's model, multiple distributed networks within the brain produce content in parallel. The content that has the biggest impact on the rest of the system achieves 'fame in the brain' and enters awareness. He uses the metaphor of fame to emphasize that there is no exact time at which any particular piece of content becomes famous, and that fame is determined afterwards.

1991	1998	2009
Daniel C. Dennett introduces the multiple drafts model of consciousness	Stanislas Dehaene and Jean-Pierre Changeux propose the global workspace theory	Raphaël Gaillard and colleagues show that conscious, but not unconscious, stimuli are widely broadcast across the brain

show, by shining a bright spotlight onto the stage to illuminate the actions of some of the actors. The spotlight of attention reveals the contents of consciousness, which change by the minute as some actors move off-stage and are replaced by others.

Although only a few of the actors are in the spotlight of attention at any one time, many continue working behind the scenes. Their activities are invisible and therefore do not enter into conscious awareness, but some can influence the activities of those who are on the stage. The spotlight of attention is surrounded by a fringe of vaguely conscious but crucial events, which can subtly alter the proceedings on the stage. In this way, unconscious information processing by the brain can influence conscious awareness. The communication between actors in the spotlight and those behind the scenes goes both ways. The spotlight of attention acts as a 'hub', which not only redirects actions from backstage to actors on the stage, but also broadcasts important information from the stage to all of the others.

SUPPORTING EVIDENCE

The global workspace model is a theoretical framework that describes the mental architecture of consciousness. Viewed within this framework, consciousness can be thought of as the mechanism by which the brain prioritizes relevant information and gives us access to it. The model successfully accounts for some of its main characteristics. The spotlight of attention explains why consciousness has a limited capacity; its constantly shifting focus explains why we experience consciousness as a stream; and the interactions between the fringe and on-stage actors explain how conscious and unconscious processing influence each other.

> **CONSCIOUSNESS ACTS AS A "BRIGHT SPOT" ON THE STAGE, DIRECTED THERE BY THE SELECTIVE "SPOTLIGHT" OF ATTENTION.**
> Bernard Baars, 1997

Although the global workspace is theoretical, there is experimental evidence to support it. In 2009, a group of French researchers had the rare opportunity of recording neuronal activity directly from the brains of conscious epileptic patients about to undergo neurosurgery. The patients were shown a series of words on a computer screen as they lay on the operating table. Some of the words were preceded and followed by a 'mask', which made them visible for just 29 milliseconds, so that the patients did not become aware of them. Others were not followed by the mask, and remained visible for longer.

The researchers used electrodes placed onto about 180 different locations on the surface of the brain, and found that the masked and unmasked words produced different patterns of brain activity. Masked words produced rapid and strong responses, predominantly in the visual cortex, but these responses decayed very quickly. By contrast, the unmasked words produced strong responses in both the visual and frontal cortices, which were followed by longer-lasting, synchronized activity throughout the brain. These findings lend support to the global workspace theory of consciousness – the researchers interpret them as meaning that the unmasked words are broadcast throughout the brain, whereas the masked words are not.

THE LATEST MODEL
Researchers have refined the original theory in an attempt to show how it might be implemented within networks of neurons. In this newer model, the global neuronal workspace consists of a set of neurons that are distributed throughout the cerebral cortex, and which communicate with each other via axons that project for long distances through the cortex. These cells accumulate competing pieces of information, and select those that are relevant to the task at hand. They then amplify the relevant stimuli and distribute (or broadcast) them to their counterpart cells in other areas of the cortex, so that they can enter the spotlight of attention and reach our conscious awareness. We are still a long way off from understanding how the brain generates consciousness, but the global workspace model is one of the most comprehensive theories yet.

The condensed idea
Consciousness is the
spotlight of attention

20 Consciousness disorders

Conscious awareness is the ability to perceive ourselves and the world around us. It is severely reduced in minimally conscious patients, and is thought to be entirely absent in vegetative patients. New methods show, however, that some vegetative patients actually retain some degree of awareness, and enable researchers to communicate with them.

Conscious awareness is an essential component of consciousness. It is generated by the coordinated activity of many different parts of the brain, particularly the cerebral cortex, which contains dozens of specialized areas for processing sensory information from the body and the outside world. Awareness also depends upon intact connections between the cortex and subcortical structures such as the thalamus, and is closely related to arousal, which is regulated by a part of the brain stem called the reticular activating system.

IMPROVING DIAGNOSIS

We know that conscious awareness is severely impaired in disorders of consciousness, such as in minimally conscious and vegetative states, but we still have no way of determining the level of awareness in patients diagnosed with these conditions, or of distinguishing between these states to diagnose them accurately. This began to change about ten years ago, with technological advances enabling doctors to diagnose such disorders more accurately. Using these new methods, researchers have shown that a significant proportion of

TIMELINE

1972	1990	2002
Bryan Jennett and Fred Plum coin the term 'vegetative state'	Terri Schiavo suffers a heart attack and falls into a persistent vegetative state	Publication of diagnostic criteria for the minimally conscious state

vegetative patients, who were thought to be completely unconscious, actually do have some level of awareness, and are also capable of communicating their thoughts, despite being unresponsive to behavioural tests.

Disorders of consciousness are most often caused by severe brain damage following a traumatic brain injury. It's estimated that there are 100,000 to 200,000 patients worldwide with these disorders, although it is currently thought that up to 40 per cent may have been given a wrong diagnosis. Each state is associated with a different outcome, but accurate diagnosis is a major challenge. For instance, patients in the minimally conscious state are typically more likely to improve than vegetative patients, but we still cannot predict who might improve, or by how much. Clinical researchers are now focused on developing methods to distinguish between these types of disorders. The ability to make accurate diagnoses could help doctors to make better predictions about which patients will recover.

> **CONSCIOUSNESS IS THE APPEARANCE OF A WORLD.**
> German philosopher
> Thomas Metzinger, 2009

Consciousness disorders include:

Coma: A coma is a state of deep unconsciousness, in which patients are unable to move, open their eyes or respond to external stimuli in any way. Comatose patients do not exhibit normal cycles of sleeping and wakefulness, and are thought to lack conscious awareness altogether. They are incapable of breathing on their own, so have to be kept alive with a respirator. People rarely remain in a coma for long periods of time – either progressing to a less severe state or dying within about two weeks.

Vegetative state: Some patients enter into a vegetative state following a short period in a coma, and those who remain in this state for more than one month with no sign of improvement are said to be in a persistent vegetative

2005	2006	2009
US federal court upholds the decision to remove Terri Schiavo's feeding tube	Adrian Owen and colleagues use fMRI to communicate with patients in the vegetative state	Researchers show that some minimally conscious and vegetative patients can learn simple associations

Terri Schiavo

The tragic case of Terri Schiavo highlights the ethical difficulties of caring for minimally conscious patients. Schiavo suffered a massive heart attack that left her severely brain damaged, and several months later she was diagnosed as being in a persistent vegetative state. Subsequently, a bitter legal dispute ensued between Schiavo's husband, who argued that she would not have wanted to live in this condition and that her artificial nutrition and hydration should be stopped, and her parents, who believed that she showed signs of awareness and should therefore be kept alive. This lasted seven years, during which time her life-support system was stopped and restarted twice, and President George W. Bush signed new legislation designed to keep her alive. It finally came to an end in 2005, when the federal court upheld the original decision to withdraw her feeding tube.

state. The sleep-wake cycle is preserved in the vegetative state, and patients appear to be awake but show no signs of conscious awareness. New methods for assessing brain function have revealed that at least one in five patients diagnosed as being in the vegetative state actually retain some level of conscious awareness.

Minimally conscious state: The minimally conscious state was recognized as a distinct consciousness disorder only recently. Minimally conscious patients show intermittent signs of awareness of themselves and their environment. They are non-communicative, but can sometimes follow simple commands, reach for and grasp objects, or smile or cry in response to emotional stimuli. Although they are more likely to improve than vegetative patients, some remain permanently in the minimally conscious state.

MAKING CONTACT

Several years ago, researchers devised a method for communicating with vegetative patients. The researchers placed patients in a scanner and asked them a series of simple questions, such as 'Do you have a brother?' They instructed the patients to imagine playing a game of tennis if they wanted to answer 'yes', or to imagine walking around their home if they wanted to answer 'no'. Each of these mental imagery tasks produces a different pattern of brain activity – the first activates the premotor cortex, which is involved in planning movements, while the second activates the hippocampus and surrounding areas, which are involved in spatial memory.

Initially, the researchers had no idea if any of the patients were even aware of the instructions and questions being put to them. To their surprise, however, some patients responded to the questions and answered them correctly, as verified later from their medical records and by family members. The researchers are now building on the early work to develop a battery of neuropsychological tests that can be administered to patients in the same way. They hope the tests will help clinicians to evaluate patients more accurately and determine the extent of their mental abilities.

The ability to communicate with such patients raises major ethical issues. Should they be asked, for example, if they wish to continue living, or if they would rather die? The researchers involved believe this is inappropriate, not least because most countries do not have euthanasia laws that permit switching off life-support systems if they say they do wish to die. Instead, they suggest that patients should be asked questions that can help caregivers make their daily life as comfortable as possible, such as 'Are you in pain', or questions about their preferences for food and entertainment.

The condensed idea
Awareness is severely impaired in consciousness disorders

21 Attention

Attention is the process by which we concentrate on certain things while ignoring others. It is the gatekeeper of awareness – we do not perceive something if we don't pay attention to it – but our capacity to pay attention is strictly limited to four items, and the brain's attentional mechanisms are highly selective.

The word 'attention' is an everyday term that can have many meanings, but in neuroscience it refers to brain mechanisms that enable us to process relevant inputs, thoughts or actions while ignoring anything irrelevant. It can be divided into 'voluntary attention', which refers to our ability to focus deliberately on something, and 'reflexive attention', the process by which something 'grabs' our attention.

Attention has interested researchers for more than a century. The great American psychologist William James recognized its key characteristics in the 1890s, noting that we can consciously control the focus of our attention, but that our capacity to do so is strictly limited. At around the same time, the German physician and physicist Hermann von Helmholtz performed early experiments to study the phenomenon.

Helmholtz looked at a screen with letters on it while illuminating a small part of it with an electric spark. By fixing his gaze on the centre of the screen, and deciding in advance which part of it to pay attention to, he could perceive the letters in that location, but not those in others. Helmholtz had stumbled upon what we now call 'covert attention', or seeing something out of the corner of

TIMELINE

1890	1894	1953
William James defines attention in his book *The Principles of Psychology*	Hermann von Helmoltz performs early studies of visual attention	Colin Cherry publishes his work on the cocktail party effect

the eye. This differs from 'overt attention', which involves shifting our gaze in a specific direction.

THE COCKTAIL PARTY EFFECT

It was not until about 50 years later that researchers began to obtain experimental evidence about the mechanisms of attention. In 1953, British psychologist Colin Cherry examined the cocktail party effect, which refers to our ability to focus on a single conversation in a noisy and confusing environment while ignoring others.

Cherry used headphones to play competing speech inputs into his subjects' two ears, and asked them to repeat immediately what they heard in one ear. He discovered that they were unable to repeat any of the words played into the other ear, and this led him to hypothesize that attending to the words from one ear resulted in better processing of those words at the expense of the words entering the other ear. Cherry also noticed, however, that subjects perceived high-priority information, such as their own name, even when it was presented to the ear not being attended to. This observation, called the 'intrusion of unattended inputs', led to the idea that the brain processes all information, regardless of whether it was attended to or ignored.

Several years later, another British psychologist, Donald Broadbent, proposed the influential 'bottleneck' theory of selective attention to explain results like these. According to Broadbent's theory, the brain's information-processing system is a channel with limited capacity, through which only a certain amount of information can pass. This channel acts as a gate: it stays open for information that is attended to, allowing it to

> ATTENTION IS ... THE TAKING POSSESSION BY THE MIND ... OF ONE OUT OF WHAT SEEM SEVERAL SIMULTANEOUS POSSIBLE OBJECTS OR TRAINS OF THOUGHT.
>
> William James, 1890

1957
Donald Broadbent proposes the bottleneck theory of attention

1999
Daniel Simons and Christopher Chabris publish their invisible gorilla study

2011
Nilli Lavie and colleagues publish the first study of inattentional deafness

enter the brain for processing, but is closed for information that is ignored. Broadbent also noted that this gating mechanism is under conscious control.

THE INVISIBLE GORILLA

Attention and awareness are closely linked, because we don't usually perceive anything that we don't consciously attend to. Although we have known that attention is highly selective for more than a century, it is only recently that the true extent of attentional selectivity has emerged. From research published in the past decade, it is now clear that focusing our attention can make us completely oblivious to sights and sounds that would otherwise be glaringly obvious.

The 'invisible gorilla experiment', first performed in 1999, is the most striking demonstration of this. Researchers asked their participants to watch a film clip of two small basketball 'teams', and instructed them to pay close attention to the players, in order to count the number of times they passed the ball to each other. Halfway through the short clip, a man wearing a gorilla suit walks on screen, stands among the six players, and beats his chest several times before walking off again. Remarkably, the researchers found that many

Sleight of hand, sleight of mind

Magicians know all too well that attention is highly selective, and are adept at manipulating the attentional focus of their audience to enhance the effects of their tricks. They know, for example, that people have a tendency to follow the gaze of others, a phenomenon known as 'joint attention'.

Magicians exploit this phenomenon by using their own eye movements to divert the audience's attention away from the hidden manoeuvres underlying their tricks. They also know that the sudden appearance of an unexpected object is very distracting, and will immediately draw the attention of the audience. Hence, pulling a rabbit out of a hat, or producing a flying dove, are commonly used tactics that divert the audience's attention, and give them further opportunities to perform hidden manoeuvres.

of the participants failed to notice the man in the gorilla suit, because they had focused their attention so intently on the actions of the players.

This failure to see something in plain sight is known as 'inattentional blindness'. In 2012, another group of researchers demonstrated the auditory equivalent, called 'inattentional deafness'. They showed their participants cross shapes on a computer screen. Each cross had one green and one blue arm, and one arm was slightly longer than the other. The participants were asked to indicate either which of the arms was blue or which arm was longer. This second task was slightly more difficult, as it involved paying closer attention in order to notice the subtle differences between the lengths of each arm.

The participants wore headphones throughout the experiment. Occasionally, a sound was played through the headphones during the task, and afterwards the participants were asked if they had heard it. The researchers found that they were far less likely to hear the sounds when they were played during the more difficult task – showing that inattentional effects can transfer across the senses. In other words, paying close attention during a visual task can make us unaware of sounds, and vice versa. This has obvious implications for daily life. For example, writing a text message while crossing the road could make you oblivious to the sound of an approaching car. But it could also have potential benefits, such as helping us to ignore distractions while concentrating on work.

The condensed idea
Attention is a narrowly focused spotlight

22 Working memory

Working memory, which is essential for human cognition, can be thought of as a mental workspace or the brain's notepad – a neural system for storing and manipulating small amounts of useful information. It is closely related to the process of attention, and has a limited capacity for storing information.

The term 'working memory' describes the brain mechanisms that temporarily store and manipulate information relevant to the task at hand. This ability enables us to plan and carry out everyday actions effectively. Working memory is essential for reading and performing mental arithmetic, for example, or for dialling a telephone number – before mobile phones became so ubiquitous, making a phone call typically involved looking up the number then repeating it several times while dialling it. Once you had dialled the number, you no longer needed the information, so you'd stop repeating it to yourself and quickly forget it.

TWO-TIERED STORAGE

An influential model, put forward by British psychologists in the late 1960s, states that memory consists of two distinct but related storage systems, named short-term and long-term memory. According to this model, memory is the result of the flow of information through three boxes, each representing a memory system. First, information from the outside world enters a sensory memory; next, the bits that we pay attention to are transferred to short-term memory. This in turn can be transferred to long-term memory if it is rehearsed. Without rehearsal, the information is lost from the short-term memory and forgotten.

TIMELINE

1885	1949	1956
Hermann Ebbinghaus examines his ability to remember strings of nonsense syllables	Donald Hebb distinguishes between short-term and long-term memory	Publication of George Miller's classic paper, 'The Magical Number Seven, Plus or Minus Two'

The neuroanatomy of working memory

Studies of brain-damaged patients and functional magnetic resonance imaging (fMRI) of healthy people reveal which brain regions are activated during different working memory tasks, while research on monkeys, which have the same working memory capacity as human beings, provides details of the underlying cellular mechanisms. All the results fit nicely with the multi-component model.

The central executive (the control centre) is associated with activity in the dorsolateral prefrontal cortex (1), with the activity increasing as more demands are made of working memory.

The phonological loop engages language areas located around the junction between the temporal and parietal lobes of the left hemisphere (2), and the visuo-spatial sketchpad with activity in the right occipital and parietal lobes (3), which process visual and spatial information, respectively, and the right frontal lobe.

Rehearsing a string of words by saying them either out loud or non-vocally activates the left frontal and temporal lobe regions involved in language, including Broca's Area (4), as well as the cerebellum (5).

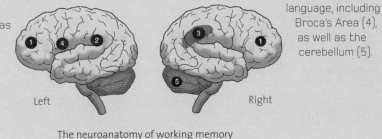

Left   Right

The neuroanatomy of working memory

This 'dual-process' model was highly influential because it explained various observations about memory function, such as why amnesic patients such as H.M. (*see page 74*) retained the ability to store small amounts of information briefly despite other, severe memory impairments. It is, however, an oversimplification, because it treats the short-term and long-term stores as each being a single mechanism, when there are actually different types of both. And so other researchers proposed the working memory model to account for the different components of the short-term store.

1968	1974	1975
Richard Atkinson and Richard Schiffrin propose the dual-process model of memory	Alan Baddeley and Graham Hitch offer the three-component model of working memory	Alan Baddeley, Neil Thomson and Mary Buchanan demonstrate the 'word-length' effect

MEMORY UPDATE

The concept of working memory updates the classical idea of a short-term memory store, but states that it consists of multiple interacting components. Working memory is a theoretical concept, but hundreds of experiments show that it is both valid and useful. There are several different models of working memory, but the best-known and most widely used is the multi-component model, first proposed in the 1970s. According to this model, working memory consists of three parts – a central executive, or control centre, that monitors and coordinates the activities of two 'slave' subsystems.

One subsystem is the phonological loop, which temporarily stores memories for words and speech-related sounds. This aspect of working memory is needed for any activity that requires remembering a verbal sequence, such as rehearsing a telephone number for long enough to dial it. The phonological loop is closely related to the brain's speech system. For example, people are far less accurate at repeating strings of similar-sounding words than dissimilar words or words with the same meaning. Word length also affects memory span – a sequence of long words is far more difficult to remember than a sequence of shorter ones, because longer words take longer to rehearse. Consequently, Welsh-speaking children are poorer at repeating lists of numbers than English children, because the Welsh words for numbers are longer.

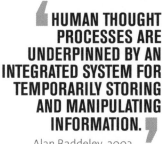

HUMAN THOUGHT PROCESSES ARE UNDERPINNED BY AN INTEGRATED SYSTEM FOR TEMPORARILY STORING AND MANIPULATING INFORMATION.

Alan Baddeley, 2003

The other subsystem, the visuo-spatial sketchpad, is concerned with temporarily storing visual information such as the colour, shape and texture of objects, and spatial information, such as the location of objects in our immediate surroundings, or the route between two destinations. Evidence that visual and spatial information is stored separately within the sketchpad comes from experiments in which tasks interfere with performance on visual but not spatial skills, or vice versa, and from brain-damaged patients who exhibit deficits in one but not the other. The ability to store and manipulate visual and spatial representations is important in fields such as architecture and engineering, and is what enabled Albert Einstein to develop his theory of general relativity.

TESTING THE LIMITS

Early research into short-term memory showed that most people's memory span is limited to about seven items. This can, however, be extended by incorporating more information into each item, a process referred to as 'chunking'. To demonstrate chunking, try to memorize this string of letters: FNBIBHBTISCV. Now try this one: NHS FBI BBC ITV. Even though both contain the same 12 letters, the second is far easier to memorize than the first, because the letters have been chunked together to form acronyms that most people are familiar with.

A classic early experiment shows that working memory is also limited in time. Participants were shown a grid containing 12 letters arranged in three rows of four, for about 50 milliseconds, and were then asked to recall as many as possible. On average, they could recall only one letter from each row. When the letters were immediately followed by an instruction to look at one of the rows, they could recall all four letters in that row. If, however, the instruction to focus on a single row came more than one second after the letters, they could recall only one item from that row, and one from each of the others. This shows that working memory is closely related to the mechanisms of attention.

The limited capacity of visuo-spatial working memory explains a curious phenomenon called change blindness, in which people fail to notice changes in a scene, such as the change in the colour or position of an object in a picture, or its disappearance. Findings such as these confirm that working memory capacity is strictly limited to four items. Attention acts as a filter that determines what enters working memory, and we only become conscious of things that we actively attend to.

> **THERE IS A FINITE SPAN OF IMMEDIATE MEMORY AND THIS SPAN IS ABOUT SEVEN ITEMS IN LENGTH.**
> George Miller, 1956

The condensed idea
Limited amounts of information stored for a limited time

23 Learning and memory

Your brain has an apparently infinite capacity for acquiring new information, and contains several distinct subsystems for learning and storing different types of data. The cellular basis of learning and memory is one of the most intensively studied topics in neuroscience, with decades of research revealing that learning changes the physical structure of the brain.

The brain gives us the remarkable ability to store apparently unlimited amounts of information, enabling us to acquire new skills, recall factual knowledge and life events, and to learn from our experiences so that we can adjust our behaviour. Learning and memory have been studied extensively for more than a century, and we know that there are several distinct types of each. The past 50 years have seen major advances in our understanding of the cellular mechanisms underlying both processes, and it is now widely believed that each involves the strengthening of connections between networks of neurons.

PUZZLED CATS AND HUNGRY DOGS

Operant conditioning is a form of learning in which behaviour is modified by its consequences. It was first studied by the American psychologist Edward Thorndike, who placed cats into puzzle boxes and watched as they tried to escape in order to reach scraps of food. They would try various means until they stumbled across a lever that opened a door.

TIMELINE

1904	1905	1949
Ivan Pavlov is awarded the Nobel Prize for his work on classical conditioning	Edward Thorndike proposes the Law of Effect	Donald Hebb proposes the mechanism of LTP in his book *The Organization of Behaviour*

Each time they were placed back in the box, they escaped more quickly than the last time, because they had learned to associate pressing the lever with a favourable outcome. On the basis of these observations, Thorndike proposed the Law of Effect, which states that any behaviour that has pleasant consequences is likely to be repeated, whereas a behaviour that has unpleasant consequences will not.

The behaviourist B.F. Skinner subsequently explained operant conditioning in more detail, using the concepts of reinforcement and punishment. Positive reinforcement strengthens a behaviour by rewarding it, whereas negative reinforcement strengthens a behaviour that removes an aversive stimulus. For example, if a rat receives food every time it presses a lever, the food positively reinforces the lever-pressing behaviour. If, on the other hand, pressing the lever stops it from receiving an electric shock, it negatively reinforces the same behaviour. Punishment has the opposite effect, and weakens behaviour by associating it with an aversive stimulus.

BEHAVIOUR IS SHAPED AND MAINTAINED BY ITS CONSEQUENCES.
B.F. Skinner, 1971

Another form of learning is classical conditioning, which was discovered accidentally by the Russian physiologist Ivan Pavlov. Pavlov was studying digestion in dogs, and noticed that they would salivate before receiving their food. In his now-famous experiments, Pavlov rang a bell while feeding the dogs. After repeated pairings of the bell and the food, the animals learned to associate the two stimuli, and would begin to salivate when they heard the bell. If, however, they heard the bell several times without being given food, the conditioned response (salivating to the bell) would extinguish, or fade away.

Both operant and classical conditioning can be used to modify human behaviour. For example, classical conditioning is the basis for aversion

1953	1966	2006
Henry Molaison (H.M.) undergoes neurosurgery to treat his epilepsy	Terje Lømo discovers LTP in the rabbit hippocampus	James McGaugh and colleagues document the first case of hyperthymestic syndrome

therapy, a treatment in which patients learn to associate undesirable behaviour with unpleasant stimuli. Alcoholics are often given emetic drugs, which make them vomit when they drink, in the hope that repeated pairings of the two will abolish their drinking behaviour.

MAKING MEMORIES

There are several different types of memory, each dependent on a distinct set of brain structures for storage and retrieval. The hippocampus is critical for memory formation and recollection, but we now know that long-term storage of memories also involves the frontal cortex, and that recalling memories becomes less dependent on the hippocampus, and more so on the frontal cortex, with time.

Declarative memory is the name for memory of facts and knowledge. It enables us to remember, for example, that London is the capital of the UK, and that George W. Bush is a former president of the USA, and is largely dependent upon the hippocampus.

Episodic memory is our memory for life events. It allows us to recall early childhood experiences, or what we had for breakfast yesterday, and involves the hippocampus and frontal cortex.

Procedural memory relates to our memory of how to do certain things, such as riding a bike, driving a car or playing a musical instrument. Learning such skills initially requires a huge amount of effort, but performing them eventually becomes automatic. Procedural memory is dependent upon the cerebellum and striatum.

Curse or blessing?

Hyperthymestic syndrome, or superior autobiographical memory, is a condition in which people literally cannot forget anything that happened to them. First described in 2006, it is apparently very rare; so far there have been only several dozen documented cases.

Individuals with hyperthymestic syndrome spend a huge amount of time thinking about their past, and have a remarkable ability to recall almost every day of their lives perfectly. For example, they can say, with great accuracy, what they did on a given day in 1982, and the events that were in the news on that day. This can interfere with daily life and can therefore be extremely detrimental. Exactly why it occurs is unclear, but research published in 2012 shows that hyperthymesia is associated with differences in brain structure. Patients with the condition have greater volumes of grey matter in brain areas linked to autobiographical memory, and increased connectivity between these areas and the frontal cortex.

Semantic memory is our memory of meanings and concepts, critical for processes such as reading for example, which depends on our ability to remember the meaning of words. Semantic memory involves the hippocampus and frontal cortex.

Spatial memory is the type of memory that records information about our environment and the relationship of objects and landmarks within it. It is largely dependent on medial temporal lobe structures surrounding the hippocampus.

WORKOUT FOR THE SYNAPSES

Both learning and memory are thought to involve the strengthening of synapses; long-term potentiation (LTP) is one mechanism by which this occurs. LTP was discovered in experiments performed on slices of rabbit brain, in which electrodes were used to stimulate axons entering the hippocampus at the same time as the cells that receive inputs from them. Stimulating the cells together causes them to fire in synchrony, and this enhances the signalling between them for days or even weeks, so that a single pulse of stimulation delivered to the axons later on evokes an enhanced response in the cells downstream. This enhanced signalling occurs by several mechanisms. Presynaptic cells can release greater amounts of neurotransmitters, for example, and postsynaptic cells can insert additional receptors into their membrane. We also know that learning and experience can lead to the formation of completely new synapses, by inducing the sprouting of dendritic spines, the tiny, mushroom-shaped structures at which synaptic transmission takes place.

The condensed idea
Learning causes physical changes to the brain

24 Mental time travel

Memory enables us to project ourselves back in time, to recall events that occurred many years ago. It is a reconstructive process, requiring us to piece together fragments of memories into a coherent and relatively accurate recollection. As such, memory also enables us to imagine future events by reconstructing bits and pieces of real memories.

Memory, wrote William Blake, enables us to 'traverse times and spaces far remote'. It allows us to perform mental time travel – we can travel back in time, to recall not only the party that we attended last Saturday night, but also those events from our remote past. But memory also serves another function: it enables us to project ourselves mentally through time in the opposite direction, and to imagine future events not yet experienced. This ability to simulate the future may be the primary function of autobiographical memory (memory for life events).

CHINESE WHISPERS

From decades of research, we know that the nature of memory is reconstructive, rather than reproductive. It does not work like a video recorder, storing events exactly as they happened. Instead, retrieving a memory involves stitching together small fragments of information into a meaningful narrative. Memory recall is therefore prone to errors, which creep in during the reconstructive process. Most of the time, our memories are accurate

TIMELINE

1932	1985	2003
Publication of Frederic Bartlett's book *Remembering*	Endel Tulving hypothesizes mental time travel, or chronesthesia	Brain scanning shows that remembering and imagining the future activate the same brain regions

enough to be reliable. Sometimes, however, the errors can be so big that they make a memory completely unreliable, as in the case of confabulation and false memories.

Early work into the reconstructive nature of memory was carried out by Frederic Bartlett, an experimental psychologist who worked at the University of Cambridge. Bartlett's ideas about memory came to him during a game of Chinese whispers, in which a story is relayed through a chain of people. Each person in the chain makes minor errors while retelling the story to the next, so that the final retelling is completely different from the original version.

Bartlett adapted the game for his experiments. In one of his studies, he asked people to read a Native American folk story called *The War of the Ghosts*, and then to recall it several times, sometimes up to a year later. He found that people invariably altered the narrative of the story upon recall. They omitted parts they believed to be irrelevant, changed the emphasis to those points they considered to be most significant, and rationalized the parts that did not make sense, to make it more comprehensible to themselves.

According to Bartlett, they did so to make the story fit in with their pre-existing framework of knowledge. In other words, the process of remembering is tainted by our own expectations and biases, which subtly alter our recollections. Bartlett published these results in a classic book called *Remembering*. He concluded that: 'One's memory of an event reflects a blend of information ... encoded at the time it occurred, plus inferences based on knowledge, expectations, beliefs and attitudes.'

> **YOU DON'T NEED MENTAL TIME TRAVEL TO REMEMBER A CHEMICAL FORMULA ... BUT YOU CAN'T REMEMBER EVENTS FROM YOUR PAST, OR ANTICIPATE YOUR FUTURE, WITHOUT IT.**
> Endel Tulving, 2003

2007

Eleanor Maguire and colleagues report that patients with amnesia have difficulty imagining the future

2007

Nicola Clayton and team show that birds can plan for the future

Birds: future planners

The ability to perform mental time travel and plan for the future was, until very recently, thought to be unique to human beings. Recent research shows, however, that other species also have these abilities.

In 2007, researchers from the University of Cambridge published a study in which Western scrub jays were kept in large cages containing three compartments. The birds were fed different kinds of food, on different feeding schedules, in each compartent, and quickly learned that they were fed less frequently in the 'no breakfast' compartment than in the others. After this training period, they were unexpectedly given pine nuts in all three compartments. The researchers found that they stored the nuts in the 'no breakfast' room more often than in the other rooms, suggesting that they did so in anticipation of being there in the future without receiving any food.

FUTURE IMPERFECT

Why is memory reconstructive, rather than a faithful re-enactment of the past? Clues began to emerge several years ago, from studies of patients with profound amnesia. In one study, published in 2007, researchers in London recruited five amnesic patients and ten healthy controls, and asked them to construct new experiences in response to short sentences describing simple, everyday scenarios, such as 'Imagine you are lying on a white sandy beach in a beautiful tropical bay' or 'Imagine that you are standing in the main hall of a museum containing many exhibits'.

The healthy control participants could perform this task very easily. The amnesic patients, however, had great difficulty imagining these new experiences, and could conjure up only fragmented, incoherent sensations that were not placed within the appropriate context. 'I can hear the sound of seagulls and of the sea,' said one of the patients. 'I can feel the grains of sand between my fingers [and] I can hear one of those ship's hooters, [but] that's about it. Really all I can see is the colour of the blue sky and the white sand ... it's like I'm kind of floating.'

All five of the amnesic patients had suffered damage to the hippocampus, a structure in the temporal lobe of the brain known to be critical for memory formation. Because of this damage, their ability to recall events from their past was severely impaired. But this study, and others that followed, showed that they were also unable to imagine the future. This suggests that remembering the past and imagining the future involve the same parts of the brain, and the same mechanisms. Brain-scanning studies of healthy people confirm this – they show that recalling past events and imagining future ones activate overlapping networks of brain structures, including the hippocampus.

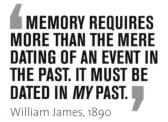

MEMORY REQUIRES MORE THAN THE MERE DATING OF AN EVENT IN THE PAST. IT MUST BE DATED IN *MY* PAST.

William James, 1890

Indeed, some researchers now argue that simulating future events is the main function of autobiographical memory, and the reason why memory evolved to be a reconstructive process instead of a reproductive one. It is because of the reconstructive nature of autobiographical memory that we can imagine events that have not yet transpired. We piece together fragments of memories of real events from our past, to generate hypothetical simulations of future events. This enables us to predict, with some degree of accuracy, how an event that we have not yet experienced will unfold, so that we can decide on the best possible course of action when the time comes.

The condensed idea
Memory enables us to recall the past and imagine the future

Memory (re)consolidation

Newly formed memories have to be stabilized, or consolidated, in order to persist for long periods of time, and recent research shows that this occurs 'offline', while we sleep. Stored memories are subsequently strengthened or 'reconsolidated', when they become temporarily unstable and can be altered or manipulated.

Memory formation is thought to involve the strengthening of synapses in a network of neurons, and activity within this network constitutes the memory 'trace'. Once a memory trace has been laid down, it has to be transferred to long-term storage, and this occurs by a process known as consolidation, whereby the memory trace is reactivated over a period of minutes to hours after it was initially formed.

Recent research shows that memories are consolidated during certain stages of sleep, and that sleep derivation can have a detrimental effect on memory function. Other work has revealed a previously unknown mechanism called reconsolidation, by which memories are retrieved from long-term storage for further strengthening. During reconsolidation, the memory trace becomes unstable, and is therefore prone to alteration.

AND SO TO BED...
The link between sleep and memory has a long history. The Roman rhetorician Quintilius noted, in the first century AD, that a good night's sleep enhances memories, and early research into the phenomenon was published

TIMELINE

1801	1924	1953
David Hartley proposes that dreaming may alter connections associated with memory	John Jenkins and Karl Dallenbach publish the first evidence that sleep promotes memory formation	Eugene Aserinsky uses electroencephalography to identify REM sleep

in the mid-1920s. The past decade has also seen a growing body of evidence which clearly shows that sleep plays an important role in the consolidation of newly formed memories.

Sleep is something of a mystery, but we have known, since the 1950s, that it consists of distinct stages, each associated with a characteristic pattern of brain waves (*see box*). As we sleep, we cycle between rapid eye movement (REM) and non-rapid eye movement (NREM) sleep, and each of these stages appears to be associated with the consolidation of a different type of memory.

Some of the evidence comes from animal studies involving spatial-navigation tasks. The hippocampus and surrounding areas contain at least three cell types that encode maps of the environment and memories for navigating it – activity in these parts can be recorded using microelectrodes

The sleep cycle

Sleep consists of five stages, each characterized by a distinct pattern of brain waves that can be detected with electroencephalography (EEG):

Stage 1: Light sleep; lasts for 5–10 minutes and is associated with large, low-frequency theta waves (*see page 165*).

Stage 2: Lasts about 20 minutes; body temperature decreases, heart rate slows down and the brain produces rapid bursts of activity called sleep spindles.

Stage 3: The transition between light and deep sleep, characterized by large, low-frequency delta waves.

Stage 4: Deep sleep; lasts about 30 minutes, and is also characterized by delta waves.

Stage 5: REM sleep; characterized by eye movements and increased breathing and brain activity. Dreaming occurs during this stage.

As we drift off to sleep, we enter stage 1 then stages 2, 3 and 4. Then we re-enter first stage 3, then stage 2, then REM sleep. We next cycle between stages 2 and 3 and REM sleep, and this occurs 4 or 5 times during an 8-hour sleep period.

1968
Early work suggesting that reactivated long-term memories are unstable

2000
Karim Nader and colleagues show that interfering with memory reconsolidation can 'erase' fearful memories in rats

2010
First report of erasing traumatic memories in human beings

implanted into the brains of freely moving animals. The hippocampus then reactivates the memory traces while the animals sleep, producing exactly the same pattern of activity. Brain-scanning studies suggest that human memory traces are reactivated during sleep, too, and this may be what causes dreams.

Most recent studies in human beings involve asking participants to memorize a list of items or learn a motor skill, then testing their memory performance the following day. Participants who get a night's sleep typically perform better on the same task than those who do not, because sleeping enhances memory consolidation. Similarly, participants who take a short nap after learning usually recall the information better than those who stay awake.

THE INTERVAL OF A SINGLE NIGHT WILL GREATLY INCREASE THE STRENGTH OF THE MEMORY.

Quintilian, c. AD 95

It follows that losing sleep could impair memory. This is also borne out in the research, which shows that sleep deprivation not only impedes performance on memory tasks, but also hampers our ability to retrieve existing memories from long-term storage. More recently, sleep deprivation has also been shown to enhance our propensity to create false memories – this has implications for interrogation methods that involve keeping suspects awake for prolonged periods of time.

(UN)STABLE FOUNDATIONS

Memory reconsolidation is the process by which memories are retrieved from long-term storage so that they can be strengthened. Soon after retrieval, though, the memory trace is rendered unstable, and can therefore be inadvertently altered or deliberately manipulated. Reconsolidation was first described in 2000, and has been proposed as the mechanism underlying the reconstructive nature of memory, because new information can be incorporated into existing memory traces during the process. The concept of reconsolidation remains controversial, however, because researchers have long thought that memories remain relatively stable after they have been consolidated.

The reconsolidation process can be exploited to 'erase' memories. More accurately, it can diminish the emotional response brought about by traumatic memories, thus making them less traumatic. In 2004, American researchers used classical conditioning to teach rats an association between a specific location in their cage and an electric shock. After training, the rats exhibited a fear response when they were returned to that location, even if they were not given a shock. The researchers then injected propranolol – a beta-blocker used to treat hypertension – into the rats' amygdala, which is involved in forming fearful memories. The drug interfered with reconsolidation of the memory, and abolished the animals' fearful response. Several years later, another research team reported that propranolol has the same effect in human beings, suggesting that it could be used to treat conditions such as post-traumatic stress disorder.

A further study, published in 2012, suggests that reconsolidation could also be manipulated to reduce ex-drug addicts' cravings and prevent them from relapsing into drug use. Ex-addicts often crave drugs when they encounter paraphernalia associated with drug use, and this often leads them to seek out and start using drugs once again. Chinese researchers showed a five-minute film clip of heroin use and drug paraphernalia to a group of heroin addicts, either ten minutes or six hours before showing them the same images again. Those who saw the video ten minutes before being re-exposed to the images showed decreased drug cravings for up to six months later. Re-exposure to the images reactivated the memories of drug use, but because this was not followed by drug use it interfered with reconsolidation of the memories associating the sight of paraphernalia with drug use. This weakened the association between paraphernalia and drug use, reducing the cravings that would normally be experienced in response to seeing the paraphernalia.

The condensed idea
Stored memories can be strengthened, altered and manipulated

26 Decision-making

Decision-making involves evaluating the risks and rewards associated with alternative choices then choosing the best possible course of action. Neuroscience is beginning to reveal the brain mechanisms underlying the decision-making process. By highlighting the role of emotions, it is also challenging the traditional view that it is a purely rational process.

We face numerous choices every day, from what to eat for breakfast to whether or not to join our colleagues in the pub after work. Ultimately, all choices are economic decisions, based on valuations of the relative risks and rewards of the available options. Neuroscience is now starting to unravel the brain mechanisms underpinning these processes, and to challenge the classical approach to how we make decisions.

NEUROBIOLOGICAL DECISION-MAKING

Brain research shows that the neurotransmitter dopamine plays an important role in how we evaluate the potential rewards of alternative options. In early work of this kind, researchers used classical conditioning to teach rodents to associate certain cues with food and water or various non-rewarding items, and then used electrodes to measure the responses of dopamine-producing neurons in the midbrain to each of the cues. These experiments showed that the cells increased their activity in response to cues that predict rewards, but not the others, suggesting that these cells use dopamine to encode reward and value. fMRI studies reveal similar mechanisms in the human brain.

TIMELINE

1836	1981	1982
Publication of *Essays on Some Unsettling Questions of Political Economy* by John Stuart Mill	Daniel Kahneman and Amos Tversky describe the 'framing effect'	Experimental economists invent the ultimatum game

More recent research suggests that two separate brain circuits are critical for the decision-making process. One, the neural system that evaluates risk and reward, comprises the dopamine-producing midbrain neurons, as well as three different parts of the frontal cortex – the ventromedial prefrontal cortex, the frontopolar cortex and the orbitofrontal cortex. The other is a network comprising the dorsolateral prefrontal cortex and anterior cingulate gyrus, which together seem to be important for cognitive control – tasks such as identifying errors and maintaining and switching focus.

Researchers examined brain imaging and behavioural data from nearly 350 patients with damage to different regions of the frontal cortex. They found that those with damage to the dorsolateral prefrontal cortex found it extremely difficult to focus their attention. In tasks involving value-based decisions, they were easily distracted by all the choices available, finding them so overwhelming that they could not decide on any single course of action. On the other hand, patients with damage to the ventromedial prefrontal cortex had difficulty assessing the risks and rewards associated with each available option. They tended to seek an immediate reward over delaying gratification, and ignored the risks involved when they perceived the potential rewards as being large.

The somatic marker hypothesis

According to the somatic marker hypothesis, emotions and feelings give rise to unconscious physiological signals, or 'markers', that modify our responses to stimuli by exerting subtle effects on the brain. This emerged from the observation that patients with ventromedial prefrontal cortex damage are not only impaired in decision-making, but also in their ability to experience emotions.

NEUROECONOMICS

Traditionally, economists regarded decision-making as a rational process in which we choose between the available options by systematically weighing

1991

Antonio Damasio proposes the somatic marker hypothesis

2010

Benedetto De Martino and colleagues examine how amygdala damage influences monetary loss aversion

The ultimatum game

This is a favourite of neuroscientists and economists who study decision-making, and usually involves the following scenario, or a variation of it. In the ultimatum game, you are told that you will be given £20 and that you can split it with a friend. You then propose how to divide the money. Your friend can either accept the proposal and take her share, or reject it, in which case neither of you receives anything. According to utility theory, you propose to give your friend the lowest possible amount – say £1 – and keep the rest for yourself. She accepts, because although you're not being very generous, she'll get something out of it rather than nothing. In reality, however, people usually propose to give away more than the minimum amount, and reject offers deemed to be too low. This is probably due to empathy, or the ability to see things from another person's perspective, further emphasizing the role of emotions in decision-making.

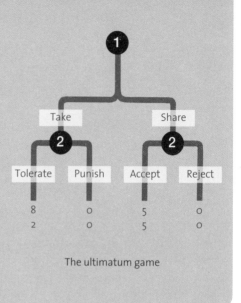

The ultimatum game

up the relative risks and rewards of each and then taking the course of action with the maximum value. This classical view, referred to as utility theory, ignores the role of intuition and emotions, however. Neuroeconomics is an emerging multidisciplinary area of research that combines methods from economics, behavioural psychology and neuroscience; it seeks to address the shortcomings in how economists view decision-making.

A classic 1981 experiment demonstrates the importance of emotions and intuition in the decision-making process. It shows how a phenomenon called framing, or presenting the same problem in different ways, affects the choices we make. In the experiment, two groups of participants were presented with a hypothetical health-risk scenario in which the USA is preparing for a disease outbreak. One group was given the choice of two programmes: programme A, in which 200 people from a group of 600 are saved, and programme B, in which there is a probability of one in three that all 600 people will be saved.

The other group was given two alternative choices: programme C, in which 400 people will die, and programme D, in which there is a one in three probability that none will die.

Statistically, programmes A and C are identical, as are programmes B and D. And yet, nearly three-quarters of participants in group one chose programme A, while a similar number of participants in group two chose programme D. The way in which the problem was presented had influenced their decision: when the outcomes were expressed in a positive light – the number of lives that would be saved – they opted for the secure choice, but when expressed negatively – in terms of the number of expected deaths – they chose the riskier option.

> **MAN ... DOES THAT BY WHICH HE MAY OBTAIN THE GREATEST AMOUNT OF NECESSARIES, CONVENIENCES, AND LUXURIES WITH THE SMALLEST QUANTITY OF LABOUR.**
>
> John Stuart Mill, 1836

Further evidence that emotions play an important role in decision-making comes from another group of patients with brain damage. It is often said that financial markets are driven by greed and fear, and this is also true for personal finances. Most of us are averse to losing money, and make financial decisions that will minimize the risk of doing so. In 2010, researchers examined two patients with a rare form of brain damage that causes the amygdala to harden and die. The amygdala is associated with the processing of emotions, especially fear. Patients with a damaged amygdala literally feel no fear, and consequently make very risky financial decisions in experimental gambling tasks.

The condensed idea
How do we choose the best course of action?

27 Reward and motivation

The brain has a dedicated reward system that motivates us to seek out the things that are essential for our survival, such as food and water. These we experience as pleasurable, and are therefore motivated to repeat the actions and behaviours that lead us to them.

Our bodies' internal needs motivate our behaviour in certain directions, leading us to specific goals that fulfil these needs. Hunger motivates us to get food; thirst motivates us to find water; and feeling cold motivates us to seek warmth. Eating and drinking are essential for our survival, and we experience them as being rewarding and pleasurable, so we have a natural urge to repeat the behaviours that enable us to obtain them. Sexual behaviour and raising children are similarly pleasurable, because they ensure our long-term survival.

Motivational states such as hunger and thirst correspond to the body's physiological states. The hypothalamus ('master gland') controls feeding behaviour and temperature regulation and coordinates brain activity with the hormonal system and the brain's reward system – consisting of structures in the midbrain, limbic system and cerebral cortex – assigns value and status to each type of reward. This determines the lengths to which we'll go in order to obtain a given reward – we are willing to allocate huge amounts of resources to high-status rewards, and less to those of lower status. Addictive drugs hijack the reward system, while motivation is affected by some psychiatric illnesses.

TIMELINE

1954	1957
James Olds and Peter Milner report that electrical stimulation of parts of the rat brain is rewarding	Dopamine is discovered by Arvid Carlsson

WELCOME TO THE PLEASURE CENTRE

The main component of the brain's reward system is a neural pathway called the mesolimbic pathway. This consists of neurons in the midbrain that produce the neurotransmitter dopamine, and which send their axons to, and form synapses with, cells in another part of the midbrain called the ventral tegmentum. Neurons in the ventral tegmental area project to a part of the limbic system called the nucleus accumbens. When these cells are

The pleasure molecule

The human brain contains about 500,000 dopamine-producing neurons, located in the midbrain. These neurons form two neural pathways, one of which is the mesolimbic pathway, or reward pathway. This contains cells that send axons from the substantia nigra to the ventral tegmentum, which in turn contains dopamine-producing neurons that project to the prefrontal cortex.

The nigrostriatal pathway contains cells that project from the substantia nigra to the corpus striatum and is involved in generating movements; it degenerates in Parkinson's disease, causing characteristic motor symptoms that can be alleviated by L-dopa, which brain cells use to make dopamine.

Dopamine is sometimes called the 'pleasure molecule' because of its role in reward, but it also encodes unpleasant experiences, and in the prefrontal cortex it is involved in attention and working memory.

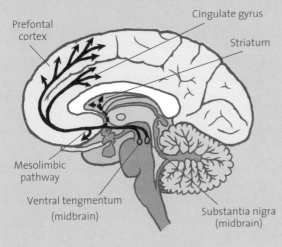

Prefontal cortex

Cingulate gyrus

Striatum

Mesolimbic pathway

Ventral tengmentum (midbrain)

Substantia nigra (midbrain)

The dopamine neurotransmitter system

2001

Robert Malenka and colleagues discover that a single dose of cocaine given to rats induces synaptic plasticity in midbrain dopamine neurons

2010

Nora Volkow and team show that ADHD involves reward system dysfunction activated, dopamine is released, leading to the experience of pleasure. The axons emanating from neurons in the nucleus accumbens form the medial forebrain bundle, which projects to a part of the frontal lobe called the orbitofrontal cortex. It is this part of the brain that assigns value to different types of rewards, and also anticipates the rewarding effects of each.

All pleasurable experiences cause midbrain neurons to release dopamine into the nucleus accumbens. This brain region is, therefore, often referred to as the 'pleasure centre'. Engaging in rewarding activities gives us pleasure and initiates learning processes that consolidate our liking of that particular goal and strengthen our memories of the situations that predict its availability. All of this reinforces the behavioural patterns that lead to obtaining the reward, making us more likely to repeat them in the future – which is a good thing for our survival.

> ❝ [MOTIVATION IS] ... ALL THOSE PUSHES AND PULLS ... THAT DEFEAT OUR LAZINESS AND MOVE US, EITHER EAGERLY OR RELUCTANTLY, INTO ACTION. ❞
>
> George Miller,
> 1962

ADDICTION AND DISEASE

Sometimes the reward system becomes skewed. Addictive drugs, for example, hijack the brain's reward mechanisms, so that addicts overvalue something that may be harmful to them at the expense of other rewards that meet their physiological and reproductive needs. Just like natural rewards such as food and sex, addictive drugs of abuse exert their pleasurable effects by enhancing dopamine transmission in the nucleus accumbens in one way or another, leading to the experience of euphoria.

Cocaine prevents cells in the nucleus accumbens from mopping up dopamine after it has been released, prolonging the effects of the transmitter at synapses in the reward pathway. Amphetamines also increase dopamine levels in the nucleus accumbens, by stimulating its release. And nicotine, the most addictive drug we know of, modulates dopamine transmission indirectly. It binds acetylcholine receptors in the nucleus accumbens, enhancing the release of dopamine.

Prolonged exposure to most addictive drugs eventually suppresses activity in the brain's reward circuitry, leading to tolerance – increasingly larger amounts of the drug are needed to obtain the same euphoric effect. Because dopamine receptor antagonists, which block the actions of dopamine by competitively binding to its receptors, can reduce cravings for some drugs, drug companies are trying to develop such compounds as treatments for addiction.

Learning plays a big role in addiction. A single dose of cocaine is sufficient to induce plasticity of the synapses in the ventral tegmental area, strengthening the pathway that releases dopamine into the nucleus accumbens. As a result, the rewarding effects of the drug are reinforced, producing the cravings associated with drug withdrawal. The sequence of actions connected with obtaining the drug is reinforced, too, leading to compulsive drug-seeking. Addiction also involves learning to associate drug-taking with specific cues, such as paraphernalia and certain situations, so that exposure to these cues reactivates the reward pathway, leading addicts to seek out the drug, and increasing the likelihood that ex-addicts will relapse into drug use.

The brain's reward system and the motivational states of the body are also altered in some psychiatric diseases. Major depressive disorder, for example, is characterized by anhedonia – an inability to derive pleasure from activities that most of us find rewarding. Consequently, depressed patients lack the motivation to engage in such activities. The reward system is also altered in children with attention-deficit hyperactivity disorder (ADHD). Children with this condition, characterized by inattention and sometimes accompanied by hyperactivity and impulsivity, need bigger incentives to modify their behaviour than others, and find it difficult to delay gratification, thus preferring to receive a small reward immediately over waiting for a larger one.

Although there has been much progress in our understanding of the brain mechanisms underlying reward and motivation, much remains to be discovered. Dopamine clearly plays a central role in these processes, and seems to be critical in how people learn to find harmful things rewarding. Further research in this area could help researchers to develop effective treatments for addiction and the various psychiatric conditions in which the reward system goes awry.

The condensed idea
What drives our behaviour?

28 Language processing

The left hemisphere of the brain was traditionally thought to contain two distinct language areas: one specialized for speech production, the other for language comprehension. This classic model is based on 19th-century studies of stroke patients with brain damage, but modern research shows that the brain's language circuits are far more complex than was once thought.

In the 1860s, a patient named Leborgne was transferred to the clinic of the French physician and anatomist Pierre Paul Broca. Ten years before being referred to Broca, Leborgne had lost the use of his right arm, and ever since then, other patients had called him 'Tan', because he was unable to say anything other than the nonsense syllable *tan*, over and over again. Leborgne died just a few days after being transferred, and when Broca examined his brain he found damage in the left frontal lobe. Subsequently, he examined other patients with the same symptoms, and noticed during autopsy that they all suffered damage in the same brain region.

Following that, the German physician Carl Wernicke described two more stroke patients. Unlike those examined by Broca, these patients could produce speech, but uttered meaningless words and sentences, and had also lost the ability to understand the spoken language of other people. Wernicke autopsied one of these patients, and discovered that he had sustained damage to another region of the brain, towards the back of the left temporal lobe.

TIMELINE

1861	1881	1949
Pierre Paul Broca presents his findings on stroke patients	Carl Wernicke describes his findings from stroke patients who cannot understand spoken language	First medical description of foreign-accent syndrome

The brain areas identified by Broca and Wernicke eventually came to bear their names. Their observations contributed to the notion of the localization of cerebral function, which states that the brain contains discrete regions specialized for particular functions (which at the time had begun to fall out of favour due to the popularity of phrenology), and to the idea that the left hemisphere is 'dominant' because it houses language functions. They also led to the classic neurobiological model of language, according to which Broca's Area is involved in speech production, and damage to it produces Broca's aphasia, or the inability to form speech. Wernicke's Area, on the other hand, came to be associated with language comprehension, and deficits in this ability are referred to as Wernicke's aphasia.

> **THE INTEGRITY OF THE THIRD FRONTAL CONVOLUTION SEEMS INDISPENSABLE TO THE EXERCISE OF THE FACULTY OF ARTICULATE LANGUAGE.**
> Pierre Paul Broca, 1861

This model was challenged at the time. Some of Broca's contemporaries noted that damage to Broca's Area did not always lead to speech deficits, while others noted that the same deficits can occur as a result of brain damage lying outside of Broca's Area. Modern brain-scanning techniques confirm this – they show that Broca and Wernicke were not entirely accurate with their anatomical descriptions, and that both language areas have functions far more complex than those originally ascribed to them. Today, neuroscientists view the classic model as being an oversimplification. Some argue that it has hindered research into the neurological basis of language, and that the terms Broca's Area and Wernicke's Area have become meaningless.

NEW IDEAS FROM OLD BRAINS

Broca's Area is a part of the brain's motor system, and as such it is responsible for controlling the muscles in the mouth and larynx that are needed for the proper articulation of speech. But modern research now shows that it is also involved in other speech-related functions. When Leborgne and another of Broca's patients, named Lelong, died, Broca preserved their brains in a Parisian

2000

Sophie Scott and colleagues identify a neural pathway involved in processing intelligible speech sounds

2007

Nina Dronkers and team use MRI to scan the brains of Broca's patients

Foreign-accent syndrome

This syndrome, a neurological condition that can occur following a stroke, causes the person to speak with what sounds like a foreign accent. It is thought to be extremely rare, with fewer than 100 documented cases since the late 1940s. In 2006, for example, British newspapers reported the case of Linda Walker, a 60-year-old woman from Newcastle, who suffered a stroke then began to speak with an accent described variously as Jamaican, French-Canadian or Italian. Foreign-accent syndrome does not actually lead to a true foreign accent, however. Instead, it is thought to occur due to subtle damage to the brain's language circuitry, which causes difficulty with the production of certain speech sounds, as well as alterations in rhythm and inclination. It seems to occur because of a disconnection between the areas that plan speech articulation and the motor areas that produce speech, and could help researchers understand how minor changes in speech-production mechanisms can alter the speech sounds produced.

museum after examining them. More than 140 years later, researchers have scanned them using sophisticated imaging techniques to find that the extent of damage was far greater than Broca had assumed at the time.

Researchers in California used high-resolution magnetic resonance imaging (MRI) to scan Leborgne's brain in 2007. The scans revealed that the most extensive damage was not in the brain region designated as Broca's Area, but in a region just in front of it. According to Broca's original reports, the damage was restricted to the surface of the brain. The scans showed, however, that it actually extended far deeper into the brain than Broca's reports suggested. Equipped with only his eyes, Broca had failed to see this. Nor had he taken into account how his patient's strokes had altered the connections between the area he identified and other parts of the brain – brain-scanning studies of stroke patients show that Broca's aphasia can occur as a result of damage to a brain structure called the insula, as well as to the basal ganglia or the white matter lying underneath the frontal lobes.

Similarly, positron emission tomography (PET)-scanning studies show that Wernicke's original anatomical boundaries are inaccurate, and that this language area contains multiple, separate subsystems, each specialized for a different aspect of language processing. These studies have identified two separate areas within what is referred to as Wernicke's Area – one involved in perceiving words and retrieving them from memory, and another activated during speech production. Wernicke's Area thus participates in functions that were traditionally ascribed only to Broca's Area. Likewise, Broca's Area is now believed to contribute to speech comprehension.

THE LIMITS OF MY LANGUAGE MEAN THE LIMITS OF MY WORLD.
Ludwig Wittgenstein, 1922

Brain-scanning studies also show that the brain region activated most frequently during speech perception is located a full 3cm (1in) farther forward than the traditional Wernicke's Area. This area was identified in 2000 by researchers in London, who also showed that it responds to intelligible, but not nonsensical, speech sounds. The area appears to be part of one neural pathway that is required for the identification of speech sounds and words. Another pathway, located farther forward and including the region traditionally referred to as Broca's Area, appears to be involved in integrating the sensory and motor aspects of speech.

The condensed idea
The brain contains multiple, complex circuits dedicated to language

29 Executive function

Executive function refers to the brain's control system, which enables us, among other things, to organize our thoughts and behaviours, prioritize and plan tasks, and make decisions. Some of these abilities – which develop throughout childhood and adolescence – accurately predict various outcomes later in life.

Also sometimes called cognitive control, 'executive function' is a term used in psychology and neuroscience to refer to a multi-component system that supervises and coordinates other high-level mental functions. It is an umbrella term used to describe a variety of processes, including attention, mental flexibility, planning, problem-solving, verbal reasoning, working memory and the ability to switch back and forth between different tasks.

Executive function emerged with the evolution of the modern mind – it is tied to the prefrontal cortex, a part of the brain far more highly developed in human beings than in our closest primate ancestors. The processes it entails are critical for guiding goal-directed actions and for the ability to handle novel situations. These are thought to be impaired in a wide range of psychiatric and neurological conditions, including Alzheimer's disease, attention deficit hyperactivity disorder (ADHD), autism, depression and schizophrenia.

We've known that the prefrontal cortex is critical for executive function since the mid-19th century, when the railroad worker Phineas Gage suffered an accident that damaged his frontal lobe and impaired his decision-making

TIMELINE

1927	1935	1962
First use of the task-switching paradigm in laboratory tests	John Ridley Stroop describes the Stroop effect	Publication of *Higher Cortical Functions in Man* by Alexander Luria

abilities (*see page 72*). Subsequently, investigations of First World War veterans with frontal lobe damage showed that they had great difficulty mastering new tasks. These observations eventually led to the idea that executive function is important for abstract, higher level thinking.

EXECUTIVE THEORIES

In the 1960s, the Soviet neuropsychologist Alexander Luria proposed that the frontal lobes are responsible for programming, monitoring and regulating our behaviour. This influential idea was more recently reformulated as the catchily named 'supervisory attentional system model'. According to this model, executive function involves multiple interacting subsystems that coordinate our goals and actions. One of these is active during routine scenarios, and monitors competing automatic responses, selecting the most appropriate and inhibiting the others. When we encounter a novel situation, the supervisory system kicks in, diverting attention as necessary to generate appropriate new responses, and providing additional inhibition and activation of automatic responses as required.

> **IN NORMAL ADULTS THE FRONTAL LOBES EXERT CONTROL OVER BEHAVIOUR IN PART AS A RESULT OF THEIR CONTROL OVER THE LEVEL OF ACTIVATION AROUSED BY DIFFERENT KINDS OF STIMULI.**
> Alexander Luria, 1979

Another influential model, proposed in 2001, builds on the idea that information processing in the brain is a competitive process. It subscribes to the prefrontal cortex the role of monitoring activity patterns in multiple brain systems – such as those involved in attention, memory, emotion and movement – and maintaining those patterns that are relevant to the present goal and the actions required to achieve it. To do so, the prefrontal cortex maps sensory inputs, internal states and motor outputs onto each other, and amplifies activity in the appropriate neural pathways to perform the current task. This is particularly important when the mappings are weak, or are continuously changing – in other words, during novel situations.

1972	**1980s**	**2001**
Walter Mischel and colleagues publish the marshmallow test in nursery school children	Tim Shallice and others propose the supervisory attentional system model	Earl Miller and Jonathan Cohen present the integrative prefrontal cortex model of executive function

The Stroop effect

The Stroop effect refers to an increased reaction time due to interference. Subjects are asked to read words of colour names or to name the colour in which the words are written. In some cases, the two stimuli are congruent (for example, the word BLACK printed in black ink), while in others they are conflicting (such as the word BLACK printed in red). When asked to name the ink colour, subjects typically take longer to respond when the stimuli are conflicting than when they are congruent. In this situation, we normally respond automatically by reading the word, so giving the correct response requires inhibiting this automatic response, which is stronger but less relevant, and selecting the weaker, more relevant one. The Stroop effect is named after John Ridley Stroop, who first described it in the 1930s, and has been used as a test of executive function ever since.

THE MARSHMALLOW TEST

Another important executive function is impulse control – the ability to inhibit automatic responses such as those that can occur in the Stroop test (*see box*), or to delay gratification. In the late 1960s, researchers at Stanford University developed the marshmallow test to examine young children's ability to delay gratification. It was performed on children aged between three and five years old, who were recruited from the university's day-care centre.

In the experiment, the researchers led each child into a room containing a table, onto which the child's preferred treat (a marshmallow, cookie or pretzel) had been placed. The children were then told they could eat it right away if they wished, but that they would be rewarded with a second treat if they could resist the temptation to eat it for 15 minutes. They were also given alternative means to distract themselves – some were given an attractive toy to play with; others were told to think of fun and pleasant thoughts while they waited; and yet others were simply left in the room without being given a toy or any instructions.

The original experiment was performed on 50 children, but since then more than 500 others have been tested. Overall, the researchers found that only a minority of the children ate the treat immediately after they had left the room. Many could resist the temptation to do so for a few minutes, and about a third could delay their gratification long enough to be awarded the second treat. Some covered their eyes so that they couldn't see the treat, while others started to kick the table leg or pull on their pigtails to try to distract themselves.

The marshmallow test began as a one-off study, but inadvertently turned into a longitudinal study, in which over a third of the original participants were followed up later in life. Numerous follow-up studies carried out in the 40 years since the original experiment show that the ability to delay gratification at a young age predicts success in later life. For example, the amount of time that a child waited before eating the treat in the original experiment was closely related to their test scores at a later date, with those who waited longer performing better in school tests.

Other follow-up studies showed that the longer a child was able to resist temptation, the higher their overall educational achievement, their sense of self-esteem, and their ability to cope with stress as an adult. On the other hand, those who opted for immediate reward in the original experiment were about 30 per cent more likely to be overweight as adults, and were more likely to abuse alcohol or drugs and to suffer from various mental illnesses.

The condensed idea
The brain's 'control system', which supervises and regulates other mental processes

30 Cell migration and axon pathfinding

Brain development is a highly dynamic process involving the mass movement of billions of nerve cells. Immature neurons migrate from their birthplace in the developing brain, then – on reaching their final destination – sprout nerve fibres, which extend towards other cells and form connections with them.

During development, stem cells in the embryonic nervous system divide to generate vast numbers of immature neurons. These then migrate en masse to form the beginnings of the brain and spinal cord. When their migration is ended, the neurons sprout two different types of nerve fibres – axons and dendrites – which grow then make synapses with other neurons to form the complex neuronal circuits found in the mature brain. Cell migration and the growth of axons and dendrites involve the same basic mechanisms; both require combinations of chemical signals, which provide directional cues that guide migrating cells and growing fibres along the right paths.

A MARVELLOUS JOURNEY

In the early stages of development, the nervous system consists of a hollow tube whose walls contain stem cells called radial glial cells. These cells divide at different rates along the length of the neural tube, with the brain forming at one end of the tube and the spinal cord at the other. Radial glial cells have a single fibre that spans the thickness of the neural tube; their cell bodies are located near the inner surface of the tube, in an area called the ventricular

TIMELINE

1868	1890	1910
Wilhelm His identifies neural crest cells	Santiago Ramón y Cajal provides the first description of the growth cone	Ross Harrison observes the movement of growth cones in a Petri dish

Migration patterns elsewhere

Brain cells are not the only cells that migrate in the body. The neural crest is a population of migratory cells that originates near the top of the neural tube. Migration of neural crest cells occurs along several distinct pathways, and involves similar mechanisms to those of axon pathfinding and cell migration in the brain, giving rise to a huge variety of neural and non-neural structures. Neural crest cells in the head region of the neural tube generate the neurons of the cranial nerves, as well as the bones and connective tissues of the face; crest cells in the trunk region of the tube give rise to primary sensory neurons and neurons of the sympathetic nervous system; and crest cells in other regions of the neural tube form, among other things, neurons in the gut, pigment-containing cells called melanocytes, and a piece of tissue that separates the aorta and the pulmonary artery.

zone. Here, they divide to produce immature neurons that migrate through the wall of the neural tube towards its outer surface.

During this so-called 'radial migration', discovered in the early 1970s, immature neurons attach themselves to the fibre of the radial glial cell that produced them, then crawl, amoeba-like, along it. In the developing brain, immature neurons migrate in successive waves to form the characteristic layers of the cerebral cortex (*see page 5*), with each subsequent wave migrating past the one before it. But the exact relationship between a cell's 'birthplace' and its final location is still unknown – cells generated in the same part of the developing brain can end up in different regions of the mature organ.

Not all cells migrate this way. The granule cells of the cerebellum, for example, are produced in a structure called the rhombic lip, located at the edge of an

1972

Pasko Rakic describes the mechanism of radial migration

2001

Takeshi Kaneko and colleagues show that radial glial cells act as stem cells in the developing brain opening in the roof of the neural tube. Immature granule neurons move away from the lip and migrate forward over the outer surface of the neural tube, before turning, then migrating down into the developing cerebellum.

THE BODY'S PIONEERS

Once a young neuron has reached its final destination, it sprouts axons and dendrites that extend towards, then begin to form connections with, other neurons. This is no mean feat – the mature brain contains something like one quadrillion connections, all of which must form correctly if it is to function properly. Furthermore, some nerve fibres extend for distances of several feet or more. Exactly how the exquisite specificity of neural connections occurs is still being actively researched, but the basic mechanisms are quite well understood. Greater knowledge of the underlying mechanisms is an exciting prospect because it may eventually enable researchers to regenerate nerve fibres that are severed in people with spinal cord injuries.

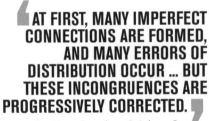

AT FIRST, MANY IMPERFECT CONNECTIONS ARE FORMED, AND MANY ERRORS OF DISTRIBUTION OCCUR ... BUT THESE INCONGRUENCES ARE PROGRESSIVELY CORRECTED.
Santiago Ramón y Cajal, 1928

From hundreds of studies carried out over the past 20 years, we know that migrating cells and growing axons detect chemical signals in their surroundings as they move through the developing nervous system. Most of these 'chemotactic' signals are proteins synthesized in specific locations, which are secreted to form a concentration gradient – in other words, their concentration is highest at their source but decreases the greater the distance from it.

Growing fibres and migrating cells detect these concentration gradients and respond to them by altering their paths. The first fibres to navigate a particular neural pathway are 'pioneers'. Those that follow can migrate along the pre-existing paths laid down by the pioneers, to form bundles of nerve fibres that connect distant brain regions.

Several distinct types of guidance cues exist, all of which contribute to help migrating cells and growing nerve fibres find, and stay on, the right path.

Repulsive cues are often secreted at the beginning of the migration pathway and, as their name suggests, help cells and growing fibres embark on their journey by pushing them away. Along the way, the correct path is delineated by permissive cues, which encourage cells and fibres to move on. The correct path is also flanked by non-permissive cues, which prevent the cells and fibres from veering in the wrong direction. As their journey reaches an end, the cells and fibres are pulled towards attractive cues secreted by cells located at their final destination.

FOLLOWING THE SIGNS

The tip of a growing axon consists of a growth cone, a cone-shaped structure approximately ten-thousandths of a millimetre wide that contains numerous finger-like protuberances called filopodia. Growth cones contain receptors for the numerous guidance cues that steer growing axons along the correct path. As a nerve fibre extends, the growth cone extends and retracts its filopodia, sniffing out the guidance cues along the way. Axon guidance cues therefore act like signposts, directing the growth cone to turn at specific locations, or continue moving straight ahead. Each type of cue initiates a distinct biochemical reaction, which leads to the reorganization of structural elements within the growth cone. For example, repulsive cues cause the growth cone to collapse on one side and to rebuild itself on the opposite side. This results in extension of the nerve fibre away from the repulsive cue.

The condensed idea
Chemical cues guide migrating neurons and growing nerve fibres

31 Cell death

The developing brain produces vast numbers of immature cells, many of which are then killed. This process – called programmed cell death – is a normal part of neural development that is under genetic control. It sculpts neuronal circuits, ensures that they contain the right numbers of cells, and determines the overall size and shape of the brain.

The brain is an incredibly complex organ containing many billions of neurons. In the womb, the developing brain produces about three times as many neurons as it actually needs, most of which are killed off before we have even been born. This process is called programmed cell death, or apoptosis, from *apo*, a Greek word meaning 'falling away from', and *ptosis*, meaning 'death', a term used to refer to the process of leaves falling from trees.

The cells that die are not defective in any way. In fact, programmed cell death is a normal practice that plays multiple important roles in the development of the brain and all of the other organs of the body. In developing limbs, for example, the immature fingers and toes are joined together by webbed tissue, an evolutionary remnant from our aquatic ancestors. As development proceeds, the cells in this tissue die off so that the limbs can develop properly.

HITTING THE TARGETS
As young nerve cells mature, they sprout fibres that extend to form connections with other neurons, muscle cells and various other 'targets'. These

target tissues produce limited amounts of chemicals called trophic factors, and neurons are dependent on these chemicals for their survival. According to the neurotrophic hypothesis, axons that grow into the same target tissue compete with each other for this limited supply of trophic factors, and programmed cell death is initiated in those cells that do not receive sufficient amounts of the chemicals.

The neurotrophic hypothesis was proposed in the early 1940s by Viktor Hamburger and Rita Levi-Montalcini, on the basis of a series of classic experiments. In the 1930s, Hamburger surgically removed limb buds – the embryonic tissues that develop into limbs – from chick embryos, and noticed that this resulted in reduced numbers of sensory and motor neurons in the spinal cord. By contrast, transplanting an extra limb bud led to an increase in the number of spinal neurons.

> **BY THE TIME I WAS BORN, MORE OF ME HAD DIED THAN SURVIVED.**
> American physician and poet Lewis Thomas, 1992

Hamburger concluded – wrongly – that target tissues provide a signal that induces immature neurons to proliferate then differentiate into sensory or motor neurons. A few years later, Levi-Montalcini repeated his experiments, but she found that the cells do not die immediately. Instead, they grow normally, and start to send fibres to their targets, only to die just before they reach them. She concluded that cell death occurs not because of lack of a signal that makes them divide and differentiate, but because of a lack a signal that promotes their growth.

In the early 1940s, Levi-Montalcini performed another series of experiments, in which she grafted mouse tumours next to developing chick embryos. This caused nerve fibres from the embryos to grow towards the tumours, supporting the idea that the tumour releases a chemical that travels to the embryo through the bloodstream. Next, she isolated sensory neurons from

1986	1991	2002
Levi-Montalcini and Cohen share the Nobel Prize for their work on NGF	Yves Alaine-Barde and colleagues discover brain-derived neurotrophic factor	Sydney Brenner, Robert Horvitz and John Sulston awarded Nobel Prize for their work on cell death genes

Factors that matter

Since the discovery of NGF in the 1950s, other growth factors have been found in the brain, all of which play important roles in the survival and growth of neurons. These include brain-derived neurotrophic factor (BDNF), which was discovered in 1991, and glial cell line-derived neurotrophic factor (GDNF), discovered in 1993.

Different populations of neurons are dependent upon different growth factors, or combinations of growth factors, for their survival. GDNF, for example, promotes the survival of many different types of neurons, including dopamine-

producing neurons of the midbrain, which die off in Parkinson's disease. GDNF, like other growth factors, is a protein, and researchers have used modern molecular biological techniques to study the gene that encodes it. This may eventually enable them to develop gene therapy for Parkinson's disease - if delivered into the brain, the GDNF gene could promote the survival of dopamine-producing neurons, to alleviate symptoms or slow down progression of the disease.

chick embryos and grew them in Petri dishes, alongside mouse tumours, and found that this caused halos of fibres to sprout from the cells and grow towards the tumours.

In the 1950s, Levi-Montalcini worked with Stanley Cohen to isolate the chemical from snake venom, which was known to induce nerve growth. Together, they showed that the chemical was a protein, and that it could induce nerve fibres to sprout when added to immature neurons grown in Petri dishes, but that this effect was abolished by the addition of an anti-serum. They named it nerve growth factor (NGF). Since then, many other neurotrophic factors have been discovered (*see box*).

DEATH OF A CELL

All cells contain several genetic pathways that trigger cellular suicide when activated. These were first discovered in roundworms and fruit flies, and are still studied mainly in these organisms, but the mechanisms of cell death are very similar in all species, and the human genome contains equivalent versions of most cell death genes.

Programmed cell death is controlled by a wide variety of signals from both outside and inside the cell, including toxins, hormones and growth factors, which can either trigger or inhibit the process. Programmed cell death also occurs following viral infection and traumatic brain injury, and in neurodegenerative diseases.

The 'core' cell death machinery is a family of executioner proteins called caspases, inactive forms of which are found in all cells. When cell death is triggered, caspases are converted to their active forms, which then act as molecular scissors, moving around the cell and destroying other proteins essential for proper cell function. This causes a series of characteristic structural changes: the cell membrane starts to bleb, or bulge; the cell's DNA becomes fragmented and the nucleus is broken apart; and finally, the whole cell breaks up into smaller fragments called apoptotic bodies.

Once a cell has died, its remains are cleared away. In the brain, this is done by housekeeping cells called microglia, which detect signals emitted by dying neurons and migrate towards them. They recognize fragments of dead cells and engulf them by a process called phagocytosis.

The condensed idea
Cell death is a normal part of brain development

32 Synaptic pruning

The elimination of certain synapses is critical for neural development as well as for proper functioning of the mature brain – the brain not only makes new synaptic connections throughout life, but also breaks them. Such 'pruning' is crucial for processes such as learning and memory, with recent research revealing a surprising mechanism by which it can occur.

The formation and maintenance of synaptic connections is essential for the proper development of the nervous system. The embryonic brain generates vast numbers of immature nerve cells, which then go on to sprout axons and dendrites that grow, branch and form an elaborate pattern of connections with other cells. At first, the brain forms more connections than it actually needs, then it cuts back superfluous or incorrect synapses to refine the developing neural pathways.

The story does not end there, however, because the brain continues to form, modify and eliminate synapses throughout our whole life span. It's now widely believed, for example, that learning and memory involve the strengthening and weakening of synaptic connections within networks of neurons, and that this also involves the elimination of synapses. Pruning of the brain's synaptic connections is, therefore, also crucial for proper functioning of the mature brain.

CUTTING WITH INTENT

Much of the research into synapse formation and elimination is performed on preparations of the neuromuscular junction (*see page 130*), which consists of the nerve terminal of a spinal motor neuron and the muscle cell with which it forms a synapse. During development, motor neuron axons grow out of the spinal cord and branch as they approach their muscle targets, forming immature synapses with many muscle cells. Each muscle cell is initially contacted by multiple axons. As development proceeds, however, most of these synapses are eliminated, so that only one synapse remains on each muscle cell.

Synaptic pruning occurs throughout the developing brain, too – one of the best-known examples is in the visual system. In the mature visual cortex, cells are arranged in so-called ocular dominance columns, which alternately receive inputs from the left and right eye. Initially, axons growing into the visual cortex form synapses rather haphazardly, so that adjacent ocular dominance columns receive inputs from both eyes. As development proceeds, many of these synapses are eliminated, to give rise to the alternating columnar pattern. This process is partly dependent upon visual experience, which fine-tunes the visual pathways and drives proper synapse formation.

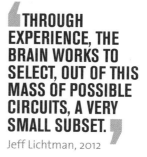

THROUGH EXPERIENCE, THE BRAIN WORKS TO SELECT, OUT OF THIS MASS OF POSSIBLE CIRCUITS, A VERY SMALL SUBSET.

Jeff Lichtman, 2012

During brain development, synaptic connections are eliminated in huge numbers. In the visual cortex of cats, for example, there is a rapid phase of synapse formation between one and five weeks after birth, followed by a phase of synapse elimination, when there is a decrease in synaptic density of approximately 40 per cent. In monkeys, the density of synapses peaks at between two and three months of age; then, from two years of age, the

2004	2010	2011
Jeff Lichtman and colleagues report shedding of retracting axons	Marie-Ève Tremblay publishes evidence that microglia prune synapses in mice	Pasko Rakic and team report that synaptic pruning continues into a human's 20s

The accessible junction

The neuromuscular junction is far more accessible than the smaller and more densely packed synapses in the grey matter of the brain, and is therefore popular among researchers investigating synapse formation and elimination. At this structure, the axons of spinal motor neurons connect to muscle cells, and use the neurotransmitter acetylcholine to elicit contraction. Initially, multiple axons connect with each muscle cell, but during development most of the axons are pruned until only one remains. This apparently occurs by a competitive mechanism, with axons vying for space at muscle cells. Researchers recently developed a laser-based technique for removing one of two axons extending to the same muscle cell. In this situation, one of the axons normally withdraws from the muscle. The researchers found, however, that destroying one of the axons caused other withdrawing axons to grow back towards the muscle to re-occupy and take over the site that has been made vacant.

density of synapses begins to decline rapidly – in the primary visual cortex, approximately 2,500 synapses are eliminated every second between the ages of two and three-and-a-half.

Similar patterns have also been observed in the human brain. At birth, the density of synapses in the human visual cortex is similar to that seen in adults, but it starts to increase rapidly at between two and four months of age, peaking at between eight and twelve months, when synapse numbers are approximately 60 per cent of those in adults.

In other parts of the brain, synapse elimination continues for much longer. The human brain reaches its full size at around ten years of age, and until recently it was thought that the organ is fully developed by then. Several years ago, however, researchers made the surprising discovery that the prefrontal cortex continues to mature well into the late 20s. In this region, synaptic pruning occurs throughout adolescence and beyond, and is necessary for the fine-tuning of the neuronal circuits involved in decision-making and other complex functions.

THE CLEANING-UP PROCESS

Various different mechanisms have been proposed to explain how unwanted synapses are eliminated. One of these, which has been observed in fruit flies, is axonal degeneration, by which unused nerve fibres wither away. In mammals, including human beings, other mechanisms have been observed. Competition for growth-promoting substances is thought to be one important mechanism: in some parts of the brain, and at the neuromuscular junction, growing axons compete for limited amounts of growth factors, with those that receive a growth signal being maintained, and those that do not withering away. Axons that do not receive a signal may retract back towards the cell body, where they are dismantled and recycled. Time-lapse footage of neuromuscular junction formation shows that axons can shed some of their materials as they retract from a muscle. Schwann cells in the vicinity then absorb the discarded material.

In the past few years, evidence has emerged that microglial cells play an important role in synaptic pruning. Microglia are the brain's immune cells, performing various housekeeping functions. They patrol the brain, continuously extending and retracting their finger-like projections to detect signs of damage. They can detect distress signals given off by injured or dying neurons, and respond by migrating to the injury site to mop up any cellular debris they find. Microglia also form the brain's first line of defence against invaders: they detect and destroy microbes that infect the organ.

It turns out that microglial cells devour unwanted synapses as if they were cellular debris or microbes. So far, this mechanism has been observed in the visual cortex and hippocampus of mice, and it could be employed in other brain regions, or throughout the whole brain. Exactly how microglial cells identify the synapses to be eliminated is still unclear, however.

The condensed idea
The brain eliminates unwanted synapses

33 Neuroplasticity

Contrary to an age-old dogma, neural circuits in the adult brain are not 'hard-wired', but change continuously throughout life as a result of our experiences, in a number of different ways. These changes are collectively referred to as 'neuroplasticity', but the term is poorly defined, and widely used and abused.

Traditionally, the adult brain was thought of as a fixed structure, which is sculpted during development then hardens as it matures, much like plaster poured into a mould. In fact, we now know that the brain changes continuously throughout life, and this is one of the most important discoveries of modern neuroscience. The changes that occur in the brain are often referred to as neuroplasticity, but as yet there is no generally accepted definition of exactly what this means. Neuroscientists use it as an umbrella term to include any of a number of physical changes that occur in the brain. For instance:

Neurogenesis: the production of new nerve cells. From decades of research, we know that several discrete regions of the brains of adult mice and rats continue to generate new neurons, and that newborn neurons make important contributions to information processing. Whether or not this also happens in the human brain is still unclear, however.

Synaptic plasticity: the strengthening or weakening of synaptic connections between neurons. It occurs in two forms: long-term potentiation (LTP) and long-term depression (LTD), which refer to the enhancement and diminishment of synaptic transmission, respectively. Synaptic plasticity is the most extensively studied and best understood type of neuroplasticity.

TIMELINE

1890	1966	1969
William James suggests that neural circuits in the adult brain are not fixed	Terje Lømo discovers long-term potentiation (LTP)	Paul Bach-y-Rita's studies of sensory substitution provide the first experimental evidence of neuroplasticity in human beings

Synaptogenesis: the formation of new connections between neurons. Neurons can sprout new dendritic spines – the tiny, finger-like projections at which synaptic transmission occurs. This has been observed directly in animals, and probably occurs in the human brain, too, but we haven't seen it because the techniques used in animals aren't applicable to people.

BRAIN TRAINING

Research shows that training of various kinds can cause physical changes in brain structure that can be detected by brain scanning. Training and practice seem to hone the neural pathways involved in the task, so that the brain becomes more efficient at performing the actions.

One of the best examples comes from studies of London taxi drivers, who spend up to four years learning the layout of the city's streets. Researchers have found that acquiring the 'knowledge' causes an increase in the volume of the hippocampus, which is involved in generating maps and storing memories of how to navigate, and that the more experienced a taxi driver

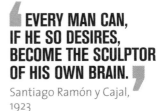

EVERY MAN CAN, IF HE SO DESIRES, BECOME THE SCULPTOR OF HIS OWN BRAIN.
Santiago Ramón y Cajal, 1923

is, the greater the density of grey matter in the hippocampus. It's not clear exactly what causes these changes, but one possible explanation is the formation of new synapses.

Similarly, spending three months learning how to juggle causes an increase in grey matter density in parts of the visual cortex involved in visuo-spatial processing. It also leads to an increase in the density of white matter tracts connecting the intraparietal sulcus, a part of the brain involved in visuo-spatial memory, to other brain areas.

More recently, researchers in London used diffusion tensor imaging (DTI) to compare the brains of karate black belts with those of novices, and found

1981	**2000**	**2004**	**2012**
David Hubel and Torsten Wiesel win the Nobel Prize for experiments on monocular deprivation in kittens	Eleanor Maguire and colleagues begin their studies of London taxi drivers	Swiss researchers show that learning to juggle changes the brain	Researchers report that karate black belts have greater white matter density

differences in the microscopic structure of the brain's white matter tracts. Black belts have a greater density of tracts connecting the cerebellum and motor cortex. This is the result of years of practice, and it enables them to pack a more powerful punch.

FOR BETTER, FOR WORSE

It is because of neuroplasticity that the brain can overcome adversity. In the 1960s, researchers discovered a process called sensory substitution, whereby blind people could be taught to 'see' with their sense of touch. Training them to use a specially built device causes the visual cortex to process touch, so that it produces a tactile image of visual information. Neuroplasticity is what enables people to recover from brain damage, too – the brain reorganizes itself to compensate for lost functions, by working around damaged areas.

In recent years, neuroplasticity has become a popular buzzword, used to explain all sorts of neurological phenomena that we still do not understand.

Eye tests

In the 1960s, researchers discovered that brain development depends upon sensory experiences in early life, and that some abnormal changes can be reversed. In a series of Nobel Prize-winning experiments, they raised newborn kittens with one eye sewed shut, and found that this significantly affected visual cortex development. The visual cortex contains alternating columns that receive inputs from the left and right eye. When sensory information from one eye was blocked, the columns devoted to that eye failed to develop, whereas those devoted to the other were much larger than normal. The researchers also found that these changes could be reversed if the unused eye was reopened within a specific, critical period of development. We now know, however, that this critical period is longer than the researchers originally thought. Today, these experiments are considered highly unethical, but they led directly to treatments for amblyopia, or 'lazy eye'.

Self-help gurus and charlatans even proclaim that their programmes induce neuroplasticity and 'rewire the brain'. When used so loosely, and without further explanation, the term becomes completely meaningless.

In fact, virtually everything we do 'rewires the brain'. Plasticity shapes the developing brain, and new experiences alter its physical structure in one way or another. Learning and memory, for example, involve the strengthening of synapses within distributed networks of neurons.

Synaptic plasticity occurs widely and constantly throughout the brain, however, for many other reasons, and it is estimated that the human brain forms about one million new connections every second of our lives. But research also suggests that the brain's ability to reorganize itself diminishes with age.

Not all neuroplasticity is beneficial. It's well known, for example, that addiction to cocaine and other drugs, including alcohol and nicotine, involves synaptic plasticity within the reward circuits that use dopamine. A single dose of a drug is enough to alter synaptic transmission, and prolonged drug use causes long-lasting changes at the synapses, leading to cravings, drug-seeking behaviour and relapse into drug use.

The condensed idea
Life experiences
reorganize the brain

34 Adolescence

Adolescence is a uniquely human phenomenon, a period of life marked by a peak in risk-taking behaviour. New research shows that the brain continues to mature throughout adolescence and into early adulthood. This results in a prolonged period of plasticity that makes teenagers highly vulnerable, but it may also have conferred an important evolutionary advantage.

Some aspects of adolescence have been recognized for millennia. Aristotle described it as something like a state of permanent intoxication, while Shakespeare encapsulated teenagers' characteristic mixture of child-like naïveté and grown-up passions in his romantic tragedy *Romeo and Juliet*. The experience of adolescence has changed significantly since then, however. Children today reach puberty earlier, and adulthood later, than they did in the past, and consequently go through a prolonged period of teenage turmoil.

TEENAGE ANGST

Stereotypical teenagers are moody, impulsive and prone to emotional outbursts. They seek new sensations and thrills and this, coupled with an apparent inability to make rational decisions, leads to risky behaviour. Teenagers also seek the approval of their peers more than that of any other group, and are therefore highly susceptibility to peer pressure. All of this is reflected in the statistics for death, disease and disability – teenagers are more likely to die in a car accident, use and abuse drugs and be sexually promiscuous than people of any other age.

TIMELINE

1590s	1904	1990s
William Shakespeare writes *Romeo and Juliet*	Publication of G. Stanley Hall's *Adolescence* establishes the field as an area of scientific research	First brain-imaging studies showing how the brain changes during adolescence

Traditionally, such behaviour was attributed to raging hormones, and adolescents do indeed experience a hormone surge that accounts for some of their actions. We now know, however, that there's much more to it than this. Brain development was long thought to be complete within the first few years of life, but recent research shows it goes on much longer than that. Although the organ has reached its full size by the age of ten, it undergoes dramatic organizational changes that continue well into early adulthood.

BLAME THE UPGRADES

The transition from childhood to adulthood involves an interaction between two neurological and psychological systems. The first is the limbic system, which is involved in emotion and motivation. Recent research shows that the nucleus accumbens, or 'reward centre', is more active in adolescents than in children or adults, and this goes some way towards explaining their risky behaviour. Contrary to popular belief, teenagers actually overestimate the risks associated with risky conduct, but because their brains are particularly sensitive to dopamine, they also overestimate how rewarding such behaviour will be.

> **ACCELERATING THE INTENSITY OF EMOTIONAL AND MOTIVATIONAL TENDENCIES ... IS, METAPHORICALLY, LIKE REVVING THE ENGINE WITHOUT A SKILLED DRIVER.**
> American psychiatrist Ronald Dahl, 2003

The other system is the prefrontal cortex, the brain's control system, which is involved in complex functions such as decision-making, long-term planning, the inhibition of impulses, and delaying gratification. Diffusion tensor imaging (DTI) shows that the brain upgrades its wiring throughout adolescence, by increasing its production of myelin, the fatty tissue that envelops nerve fibres and facilitates the conduction of nervous impulses. Myelination progresses in a wave that gradually sweeps through the brain, starting from the back of the organ to the front.

2011

Laurence Steinberg and colleagues show that peers increase adolescents' risk-taking by enhancing activity in the brain's reward circuitry

2012

US Supreme Court bars mandatory life sentences for juveniles convicted of murder

Bowing to peer pressure

Laurence Steinberg of Temple University has developed a video game to test how adolescents weigh up relative risks and rewards. The game involves taking control of a car and driving it across a virtual town as quickly as possible. Participants encounter several sets of traffic lights on the journey, some of which turn amber as they are approached. This forces the participants to make a quick decision about whether to lose time by stopping, or save time and earn points by speeding up and driving straight through. Steinberg has found that when teenagers play the game alone, they risk driving through the lights about as frequently as adults. When a friend is watching them play the game, however, they run through the traffic lights roughly twice as often. This demonstrates the influence that peers have on teenage behaviour – teenagers experience approval by their peers as being highly rewarding, and are therefore more willing to take risks in order to gain it.

Thus, areas nearer the back of the brain, such as those involved in vision, are upgraded relatively early on, while the prefrontal cortex, located behind the eyes, does not reach full maturity until the late 20s or early 30s. At the same time, connections throughout the cerebral cortex are continuously being refined by experience, and this synaptic pruning also continues into early adulthood. Consequently, teenagers are prone to making bad decisions and taking big risks, but this does not usually last.

Teenagers learn from their mistakes through trial and error, and this influences the development of the prefrontal cortex, so that their control system improves with time. This is facilitated by the ongoing processes of myelination and synapse formation, which lead to faster connections and increase the brain's networking to improve its control system. And the creation of stronger connections between the prefrontal cortex and hippocampus mean that memories of past experiences are increasingly integrated into the decision-making process.

DRIVING WITHOUT DUE CARE

Some researchers liken the adolescent brain to a car with a reckless driver who has their foot pressed down on the accelerator but still has not mastered control of the vehicle. During this stage of development, risk-taking behaviour is at its peak. Teenagers often put themselves into dangerous situations and, inevitably, are highly vulnerable. Viewed through the lens of evolution, however, the risk-taking behaviour that characterizes adolescence can be seen as a valuable adaptation.

Teenagers take risks because they seek novelty and sensation, and this primes them to move away from familiar ground and pursue new experiences and explore new territory. It thus prepares them for leaving the family home, a safe environment built by their parents, and to venture out into the world that awaits them. Delayed maturation thus makes the adolescent brain more flexible to enable it to learn from experience during this crucial stage of life. Indeed, were it not for the teenage antics of our ancestors, humankind may not have spread to every corner of the globe.

THE YOUNG ARE HEATED BY NATURE AS DRUNKEN MEN BY WINE.
Aristotle, c. 350 BC

ADOLESCENTS AND THE LAW

The surprising new finding that the prefrontal cortex does not reach full maturity until early adulthood has led some researchers to argue that we need to rethink how adolescents are treated within the legal system. Until recently, teenagers found guilty of murder in the United States were usually given a death sentence, or mandatory life imprisonment without parole. In October 2012, however, the US Supreme Court barred mandatory life sentences for juveniles convicted of murder. This was based on the growing body of evidence showing that teenagers' decision-making abilities are not yet fully developed – but it does not mean that teenagers should not be held responsible, and punished, for their actions.

The condensed idea
Teenage risk-taking behaviour may be an evolutionary adaptation

35 Stress and the brain

Prolonged exposure to stress during early childhood and adolescence has negative and long-lasting effects on the brain and behaviour. Early life stress disrupts the development of brain circuits and impacts negatively on mental function, increasing the risk of mental illness later in life. Some of these effects are probably reversible, however, and this has important implications for child-rearing and social policy.

The ability to respond to stressful or threatening situations is critical for survival, but we now know, from numerous animal and human studies, that prolonged exposure to stress has toxic effects on the brain. The brain is particularly sensitive to stress during early childhood, adolescence and old age, because it normally undergoes dramatic changes during these periods of life. Exactly how stress affects the brain depends on when we are exposed to it, and the length of time of our exposure.

Prolonged stress, particularly during early childhood but also in adolescence, can disrupt the development of brain circuitry and have long-lasting, detrimental effects on behaviour. Research shows that repeated exposure to stressful situations such as neglect, child abuse and poverty can stunt brain growth, leading to persistent and negative effects on mental functions later in life, as well as an increased risk of mental health problems.

TIMELINE

1985	1999	2005
Bruce McEwan and colleagues report that glucocorticoids kill neurons in the rat hippocampus	Early evidence that severe childhood abuse is associated with reduced brain volume	Martha Farah and colleagues link socio-economic status to academic performance

Early experiments showed that exposing rats to glucocorticoid stress hormones for prolonged periods destroys nerve cells in the hippocampus, a brain structure critical for memory. Hundreds of studies show that keeping rodents in an impoverished environment has significant negative effects on hippocampus development and function. More recently, it has also been found that the quality of a female rat's maternal care has direct effects on glucocorticoid receptors in the hippocampus of her offspring, leading to life-long 'programming' of their stress response. Much of the work on animals appears to be directly applicable to human beings, and this has numerous, wide-ranging implications.

The stress response

Stress causes a coordinated response involving the autonomic nervous system, the immune system and the hormonal system, which is controlled by the hypothalamus-pituitary-adrenal axis. Exposure to stress causes the hypothalamus to release corticotropin-releasing hormone (CRH) and vasopressin, triggering the release of adrenocorticotropic hormone (ACTH) from the pituitary gland. This, in turn, causes the release of adrenaline, noradrenaline and glucocorticoid hormones from the adrenal glands, which exert effects on many parts of the body.

Positive stress occurs as a result of a brief stressful experience, such as a child's first day at nursery. This leads to an acute response involving increased heart rate and altered hormone levels, and can be coped with easily. Tolerable stress also occurs following experiences such as divorce or death of a loved one, and children can overcome this with the help of a caring adult. Toxic stress follows adverse events that last for days, weeks or months, and can cause long-lasting changes to the brain.

FUTURE BRAIN HEALTH

Socio-economic status in early life is an important determinant of future health. We've known for many years that wealthier people tend to be healthier and live longer than poorer individuals, but only in the past few years have we begun to understand that poverty can have a direct and powerful impact on brain development and patterns of behaviour in later life. Children raised in poverty grow up in

2011

Seth Pollak and team link low socio-economic status in early childhood to reduced hippocampal volume

2012

Martin Teicher and colleagues show that severe child abuse is associated with reduced hippocampal volume impoverished environments and often lack any form of mental stimulation, and this places huge demands on them, making early life extremely stressful. Consequently, their brains fail to develop properly, setting them on course for poorer academic performance, dim career prospects and an increased risk of a variety of medical and mental health issues.

A spate of studies carried out since 2005 show that children from a low-income background lag way behind those from a middle-income background on almost every measure of cognitive development. Poorer children also consistently perform worse on tests of language, memory and visuo-spatial skills, and research shows that these differences are related to alterations in both the structure and the function of the brain.

IT IS EASIER TO BUILD STRONG CHILDREN THAN TO REPAIR BROKEN MEN.
American social reformer
Frederick Douglass, 1849

One recent study demonstrated that prefrontal cortex activity associated with executive function and working memory is altered in children from a poor background, and that this is associated with attention deficits. Another compared more than 300 children from across the socio-economic spectrum, and found that those from low-income households had lower grey matter density in the hippocampus than those from middle- or high-income households. The latest research shows that severe child abuse is also linked to a reduced volume of grey matter in the hippocampus.

The ramifications of poverty, neglect and abuse appear to depend on timing, because different parts of the brain become sensitive to the effects of stress at different stages of development. The hippocampus is particularly vulnerable to stress at between three and five years of age, whereas the prefrontal cortex becomes sensitized at 14–16 years of age. Many of the effects are initially invisible, becoming apparent only during puberty or even later. Regardless of the cause, prolonged exposure to stress in early life increases the subsequent risk of depression, psychosis, post-traumatic stress disorder, disorders of impulse control and personality, and alcohol and substance abuse, among others.

ALL IS NOT LOST

Importantly, the research also shows that at least some of the toxic effects of stress can be reversed if early action is taken. The brain development of rodents reared in isolated and impoverished environments can be rescued by environmental enrichment – the stress responses of rat pups raised by neglectful mothers can be normalized by transferring them to females who lick and groom them more often.

This, too, seems to be directly applicable to human beings, and suggests that various interventions could alleviate the negative impact of early life stress. The adverse effects of raising a child in poverty could be at least partially reversed or slowed down by providing as much mental stimulation and enrichment as possible. One recent study of more than 1,200 middle-aged Americans showed that warm, caring mothers can partly counter the negative health effects of poverty.

Such interventions could be implemented at many different levels, from parenting behaviour to educational and social policy. But timing is crucial – and interventions would probably be more beneficial the earlier they are adopted. This is, of course, much easier said than done. And there are other factors, such as genetic predispositions, that are completely out of our control.

The condensed idea
Prolonged stress is toxic to the brain

36 The ageing brain

As we get older, the brain slowly deteriorates, and this is usually associated with a decline in mental abilities. But new research suggests that the brain undergoes functional changes that can compensate for age-related degeneration, and scientists are now beginning to gain a better understanding of how certain lifestyle choices can offset the ravages of old age.

We all fear the downward slide of old age, and are familiar with the pessimistic picture of what happens in our twilight years. It's commonly thought that the brain is at its peak in the late 20s and that it's all downhill from there on – that the brain begins to wane irreversibly, leading to impaired mental faculties and, possibly, to senile dementia. Happily, this picture is not entirely accurate.

The brain does wither away slowly from the age of about 50, and although young adults consistently outperform older people on various lab tests, this does not translate to everyday experience. There are significant differences in how ageing affects people: many do experience cognitive decline as they get older, but many others continue to function normally well into their later years, and most of us know someone who remains physically fit and mentally sharp in their 80s or beyond.

Brain-scanning technologies are now providing a window into how the brain changes as we get older and a new, somewhat surprising, picture of the

TIMELINE

380 BC	**1950s**
Publication of Plato's *Republic*	Harold Brody publishes early studies of how ageing affects the brain

ageing brain has emerged in the past decade or so. Research seems to show that the brain undergoes functional changes to offset the decline that occurs with age, and scientists have even identified some people whose brains seem to be completely immune to the ravages of old age (*see page 146*).

HANDS UP FOR EXPERIENCE

Ageing typically involves a slowdown in various mental abilities, the best studied of which is memory. Older people often experience lapses in episodic memory, forgetting details such as where they parked their car, and this may be due to deficits in encoding, storage or retrieval of memories. In laboratory tests, they show significant impairments in tasks that involve switching their attention rapidly from one thing to another, and in working-memory tasks involving storing and manipulating information for short periods of time.

> **MAN WHEN HE GROWS OLD … CAN NO MORE LEARN MUCH THAN HE CAN RUN MUCH; YOUTH IS THE TIME FOR ANY EXTRAORDINARY TOIL.**
>
> Plato, c. 350 BC

On the other hand, older people usually do not experience problems with semantic (conceptual) memory, and their knowledge of the world often exceeds that of younger adults. It has also been reported that older adults have more empathy, and enjoy a higher level of emotional well-being, than younger people.

AGE-RELATED CHANGE

Ageing involves various changes in brain structure, but exactly how these are linked to cognitive function is still unclear. The most obvious structural change is a small but significant reduction in grey matter density. As we get older, the grey matter shrinks, particularly in the frontal cortex, hippocampus, caudate nucleus and cerebellum, leading to a 10 per cent or so reduction in total brain size between 20 and 90 years of age. This is associated with the death of cells in the cerebral cortex. According to one estimate, about 9.5 per cent of cortical cells die during this time – the equivalent of 85,000 neurons

2003

Bente Pakkenberg and colleagues estimate the number of neocortical neurons that die each day

2012

Theresa Harrison and colleagues identify SuperAgers, whose brains are apparently immune to ageing

SuperAgers

American researchers recently identified a small group of people in their 80s whose brains appear to be resistant to the effects of ageing. In lab tests involving memorizing lists of words, these individuals – named 'SuperAgers' by the researchers – outperformed other healthy people of a similar age and matched the performance of healthy adults aged between 50 and 65.

Structural MRI scans further revealed that their brains do not undergo the deterioration normally associated with ageing. Their neocortices were just as thick as those of the younger adults, and the overall volume of their brains was about the same, too. One region in particular – the anterior cingulate gyrus – was even thicker in the SuperAgers than in the healthy younger adults. These findings show that age-related brain deterioration and mental decline are not inevitable. Further studies of these individuals may provide clues about how to prevent or reduce the impairments in mental faculties normally associated with ageing.

per day, or one every second – leading to a thinning of the cortex and a reduction in its overall weight and surface area.

Brain-scanning studies show that ageing involves reductions in white matter density, too. This decrease is widespread, but is particularly evident in the white matter tracts lying underneath the frontal, temporal and parietal lobes. White matter in the corpus callosum, the massive bundle of nerve fibres connecting the two hemispheres of the brain, also degenerates with age. These changes seem to be better correlated with the slow decline of mental functions than the changes in grey matter, and may be linked to the reduction in the speed of information processing.

As we age, the brain undergoes various chemical changes, too – many studies show, for example, that we produce less of the neurotransmitter dopamine with age. The number of dopamine receptors also decreases throughout the brain as we get older, and this may be linked to the impairments in attention, memory and movement that many older adults experience.

With age we may also develop amyloid plaques and neurofibrillary tangles in the brain tissue. These structures are the pathological hallmarks of Alzheimer's disease, and although their appearance is a normal part of ageing, the disease develops only in some people. The reasons for this are unclear, but it suggests that Alzheimer's is the result of an abnormal or accelerated ageing process.

COMPENSATION CLAIMS

The brain retains an ability to change throughout life – a phenomenon known as neuroplasticity – but all the evidence implies that its capacity to do so decreases with age. Nevertheless, recent research suggests that the brain undergoes functional changes that compensate for age-related deterioration. Numerous brain-scanning studies show that certain regions of the brain are more active in older adults than in younger people during a wide variety of processes, including tasks involving motor control, and autobiographical, episodic and working memory.

Why some people are affected more by ageing than others is unclear, but it is likely that there are genetic variants that make us more or less susceptible to the effects. It seems that certain lifestyle choices – such as education, regular exercise, a healthy diet, good sleeping patterns and even socializing – can promote healthy ageing, and may offset any genetic predispositions to some extent. In recent years, the popularity of brain-training software has increased hugely. Manufacturers claim that these programs can combat age-related mental decline and reduce the risk of Alzheimer's. Such programs may improve performance on the skills they involve, but there's still no evidence that these benefits transfer to mental abilities in general.

The condensed idea
The brain compensates for age-related deterioration

37 Neurodegeneration

Neurodegenerative diseases are progressive, age-related conditions that share a common pathological mechanism. These diseases affect many millions of people around the world, and may become more prevalent in years to come, as the population of the Western world ages. As such, they represent a major health burden with enormous cost implications.

N eurodegenerative diseases cause the death of specific groups of neurons in the central nervous system. They can be loosely classified into two categories:

Dementias:	**Movement disorders:**
• Alzheimer's disease (the most common and best-known form of dementia) • Fronto-temporal lobar dementia • Vascular dementia • Pick's disease	• Parkinson's disease • Huntington's disease • Motor neuron diseases (such as amyotrophic lateral sclerosis, ALS) • Spinocerebellar ataxias

Transmissible spongiform encephalopathies, or prion diseases, are another group of neurodegenerative conditions. These cause both dementia and movement problems, and include:

TIMELINE

1872	1906	1937
George Huntington provides the first thorough medical description of Huntington's disease	Alois Alzheimer describes the first case of the disease that bears his name	Transmissibility of scrapie demonstrated in a population of Scottish sheep

- Bovine spongiform encephalopathy (BSE, or 'Mad Cow Disease')
- Variant Creutzfeldt-Jakob disease (vCJD)
- Gerstmann-Sträussler-Scheinker syndrome
- Fatal familial insomnia
- Kuru (*see box*)
- Scrapie (in sheep)

THE PRION HYPOTHESIS

Prion diseases are extremely rare conditions, and remained in relative obscurity until the late 1980s, when they came into the public eye following an epidemic of 'mad cow' disease that ravaged British cattle stocks. Subsequently, 156 people died of variant Creutzfeldt-Jakob disease, apparently because they ate beef from infected cows. Most diseases are caused by microbes, but prion diseases are unique: according to the prion hypothesis, they are caused by an abnormal form of a nerve cell protein that can be transmitted both between organisms of the same species and also across species.

The term 'prion' means proteinaceous infectious particle, and describes this unique mode of transmission. The prion protein is found in all neurons,

Cannibalism and the 'shaking death'

Kuru is a prion disease that was discovered in the 1950s among the South Fore peoples of Papua New Guinea, who transmitted it by practising ritualistic mortuary cannibalism. When a tribal member died, female relatives customarily dismembered and ate the body, including the nervous system. Kuru victims in particular were prized as rich sources of food, because the disease caused a build-up of fatty tissue that resembled pork. The term 'kuru' means shaking death in the Fore language, and describes the symptoms – the disease mainly affects the cerebellum, causing unsteady gait and tremors. During the 1950s, an outbreak claimed the lives of about 1,000 Fore tribe members, before cannibalism was outlawed by the Australian government. About five years ago, however, researchers returned to Papua New Guinea and identified 11 cases. They suggested that kuru has an exceptionally long incubation period, raising concerns that an epidemic of vCJD may occur in the UK following the BSE crisis of the late 1980s.

1982	1986	1997
Stanley Prusiner coins the term 'prion'	BSE ('mad cow' disease) appears in the UK	Prusiner awarded the Nobel Prize for his work on prions

Cause or effect?

In some neurodegenerative diseases, protein clumps are a direct cause of the symptoms, but in others the relationship is not so clear. It's widely assumed that amyloid-beta and tau deposits cause Alzheimer's, and that drugs which block the process can prevent or slow the disease. But this has yet to be clearly established, and it's just as likely that the clumps may be an effect of the disease rather than the cause.

but its normal function is still unknown, although its location on the cell membrane suggests that it is involved in cell-to-cell signalling. Mutations in the prion gene cause the protein to fold into an abnormal shape, and the abnormally folded molecules accumulate to form insoluble clumps that are toxic to neurons. These clumps then break up into smaller fragments, which act as 'seeds' that can spread and cause normal prion protein molecules to adopt the abnormal configuration.

MISFOLDED PROTEINS

Nearly every known neurodegenerative disease involves a pathological, prion-like mechanism, whereby inherited or spontaneous genetic mutations cause misfolded proteins to be deposited as insoluble clumps or fibres in or around neurons, interfering with them in some way. The type of protein involved, and the distribution and exact effects of the clumps, vary depending on the disease. Some neurodegenerative diseases involve more than one misfolded protein, and in many cases, protein deposits begin long before any symptoms appear.

Alzheimer's disease, for example, is characterized by the deposition of amyloid-beta protein, which forms structures called plaques in the spaces between neurons, and tau protein, which forms neurofibrillary tangles inside cells. Such abnormalities spread from cell to cell in much the same way as a virus. Cell death causes the brain to shrink, beginning normally in the hippocampus and causing problems with memory and spatial navigation. This shrinkage can be detected with brain imaging long before symptoms appear, but the clumps can be detected only under the microscope, so a definitive diagnosis is usually made after the patient has died and their brain has been examined.

Similarly, Parkinson's disease is characterized by the accumulation of misfolded alpha-synuclein protein, which forms structures called Lewy bodies inside neurons, and by the death of dopamine-producing neurons in the midbrain. Huntington's disease, on the other hand, involves a mutated protein called Huntingtin, which accumulates inside the nucleus of neurons.

Normally, misfolded proteins and other cellular debris are targeted for destruction, either by microglia, the brain's housekeeping cells, or by a biochemical reaction (called the ubiquitin-lysosome pathway) that acts like a cellular dustbin. There is a growing body of evidence that these mechanisms are faulty in neurodegenerative diseases, and this may be why misfolded proteins accumulate instead of being cleared away.

A LOOMING EPIDEMIC?

Inherited mutations can cause severe, early-onset cases of neurodegenerative diseases, but most cases are sporadic, and age is the single biggest risk factor. For example, the likelihood of developing Alzheimer's doubles for every five years over the age of 65, and after the age of 85 the risk reaches about 50 per cent. The reasons for this are unclear, but some researchers suggest that it may be because neurodegenerative diseases are caused by an accelerated rate of the normal ageing process.

The population of the Western world is ageing, partly because of the dramatic increase in life expectancy in the past century, but also because of a low birth rate. Consequently, more than half of the population in western Europe and north America are over the age of 50. This number will continue to grow as the baby-boomer generation reaches retirement in the coming decades, leading some to predict a major increase in the number of people who will develop a neurodegenerative disease. Alzheimer's, the most common neurodegenerative condition, currently affects an estimated 5.4 million people in America alone, and nearly 500,000 in the UK. This has been projected to double or treble by the year 2050.

THE AUTOPSY REVEALS CHANGES THAT REPRESENT THE MOST SERIOUS FORM OF SENILE DEMENTIA ... PECULIAR DEEPLY STAINED FIBRILLARY BUNDLES.

Early description of Alzheimer's pathology by German psychiatrist Emil Kraeplin, 1910

The condensed idea
All neurodegenerative conditions involve a common pathological mechanism

38 Adult neurogenesis

For much of the past 100 years, it was widely believed that the adult brain does not produce new cells. Such thinking began to change in the 1960s, with studies showing that some species of mammals do generate new brain cells throughout adulthood. The jury is still out, however, on whether this holds true for the human brain.

The growing brain produces vast numbers of cells in a process called neurogenesis, but for a long time it was thought that the production of new cells was restricted to the developmental period. Once a central dogma of neuroscience, the theory was challenged in the 1960s, when a series of studies showed that certain regions in the brains of rats and mice continue to produce new cells long after development has ended. These studies were initially dismissed, but subsequent research in songbirds, then monkeys, confirmed the early findings.

The discovery of adult neurogenesis is often touted as the most important discovery of modern neuroscience. It is now widely believed that the human brain produces new cells throughout life, an idea that has revolutionized the way we think about healthy and diseased brains. Scientists and the general public alike are enamoured by the idea that the adult human brain can produce new cells, because it offers hope that the brain can repair itself after injury or disease. Animal studies also show that physical and mental exercise can boost the growth of new brain cells and this is now thought to

TIMELINE

1962	1980	1992	1998
First evidence of adult neurogenesis in rats	Fernando Nottebohm reports adult neurogenesis in songbirds	Researchers isolate stem cells from the adult mouse brain	Fred Gage and colleagues find evidence of adult neurogenesis in humans

Attuned to the seasons

A series of classic studies performed in the 1980s showed that the size of the songbird brain changes with the seasons. Every year, just before mating season, the song-producing region of the male canary brain produces new cells and increases in size, enabling the male to learn new songs so that it can serenade potential mates. At the end of the mating season, the cells die off, and the song-producing region shrinks. This cycle of regeneration and degeneration is repeated on an annual basis, producing seasonal fluctuations in the size of the song-producing region. This was significant because, while early evidence of adult neurogenesis in rodents was largely ignored, the work on the canary brain led to wide acceptance of the phenomenon, and overturned the long-standing dogma that the adult vertebrate brain does not produce new cells.

apply to human beings, too; many people believe that exercise can reduce the risk of age-related cognitive decline, as well as conditions such as Alzheimer's disease and depression, by inducing adult neurogenesis. Today, this is an area of intensive research, but the question of whether or not the adult human brain produces new cells is still being hotly debated.

FROM MICE ...

Although the field of adult neurogenesis got off to a shaky start, there is now irrefutable evidence that the brains of rodents contain at least two small populations of stem cells which retain the ability to produce new neurons throughout life, and that newborn cells play important functional roles (*see also Chapter 43: Neural stem cells*).

One of these areas produces young neurons that migrate a short distance into the hippocampus. Here, they integrate into the existing neuronal circuits

1999
Elizabeth Gould and team discover adult neurogenesis in the monkey hippocampus

2006
Jonas Frisén and associates fail to find neurogenesis in the adult human cortex

2007
Maurice Curtis and colleagues find migrating cells in the adult human brain

2012
Frisén and team fail to find newborn cells in the adult human olfactory bulb

and contribute to information processing. Researchers have used genetic engineering to create mice that do not produce these new hippocampal neurons. Studies show that blocking adult neurogenesis in this way causes severe memory impairments – the animals have difficulty forming new memories, and their spatial navigation abilities are diminished.

Cells produced in the other area migrate a much greater distance, to the olfactory bulb at the frontmost tip of the brain, where they play important roles in the processing of smell information. When the production of these cells is interfered with, the mice are unable to form new memories of smells.

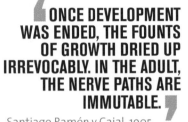

ONCE DEVELOPMENT WAS ENDED, THE FOUNTS OF GROWTH DRIED UP IRREVOCABLY. IN THE ADULT, THE NERVE PATHS ARE IMMUTABLE.

Santiago Ramón y Cajal, 1905

These animal studies also show that the number of new cells produced decreases as the animals get older. They indicate, too, that physical exercise promotes the growth of new brain cells, as does raising the animals in an enriched environment containing objects they can explore and play with. Plus, there is some evidence that the mouse brains contain stem cells which divide in response to brain injury, to produce immature neurons that then migrate to the injury site.

... TO MONKEYS AND MEN

The big question is: can these animal findings be extrapolated to human beings? It is often taken for granted that they can, but in fact the evidence for adult neurogenesis in the human brain is very thin on the ground. Throughout the 1980s and 1990s, several groups of researchers looked for new cells in the brains of monkeys, which are more closely related to our own species than rodents. These studies provided conflicting evidence, however – some researchers claimed to have identified newborn neurons in the hippocampus and cortex, while others failed to find any.

Studies of adult neurogenesis in the human brain are also conflicting. A breakthrough came in 1998, when researchers examined the brains of five patients who had died of cancer and found small numbers of stem cells in the hippocampus, which apparently retain the ability to divide and produce immature neurons. They concluded that the human hippocampus, like that of mice, produces new cells throughout life. Importantly, though, they cautioned that the study did not provide any proof that these new cells are functional.

Since then, several groups of researchers have isolated stem cells from various regions of the human brain, and found that they can produce immature brain cells when grown in Petri dishes. Another study, published in 2007, showed that the human brain contains large numbers of cells migrating to the olfactory bulb, although this finding has not been replicated. Indeed, subsequent studies showed that large numbers of neurons continue to migrate to the olfactory bulb in the brains of young infants, but that the adult brain contains far fewer migrating cells, if any.

> EVERYONE WANTS TO BELIEVE THAT WE CAN REPAIR DAMAGED BRAINS, BUT THERE'S PRECIOUS LITTLE EVIDENCE FOR IT.
> British neurobiologist Andrew Lumsden, 2011

Overall, the evidence suggests that the human brain continues to produce large numbers of new cells for a short time after birth, but that this process declines very rapidly. Evidently, the brain does contain stem cells that can continue to divide into old age – but the critical point is that we still do not know if new cells are produced in large enough numbers to be of any functional significance. Thus, the jury is still out.

The condensed idea
Does the adult human brain produce new cells?

39 Epigenetics

Epigenetics bridges the gap between nature and nurture, by showing how genes interact with the environment. It explains how life experiences can alter gene activity without changing the DNA sequence. Epigenetics plays an important role in brain function and may lead to new treatments for neurological diseases.

How is it that neurons are completely different from skin cells or lung cells, despite the fact that all three carry exactly the same DNA? And why is it that identical twins behave differently, despite having the same genetic code? The answer lies in epigenetics, an emerging field that has radically changed biology in the past ten years.

Epigenetics is the study of heritable changes in genetic activity that occur without modifying the DNA sequence. It explains how life experiences and environmental changes alter the activity of genes, and how these alterations can be transmitted to the next generation. Our understanding of epigenetics is still in its infancy, but we already know that it is involved in virtually every aspect of brain function.

LAMARCK VERSUS DARWIN

About 200 years ago, Jean-Baptiste Lamarck proposed a theory for the origins of life on Earth. He believed that organisms changed during their lifetime to adapt to their environment, and that these changes were passed on to their offspring. A giraffe's neck, for example, would gradually extend as it tried to reach higher tree branches, so that its offspring inherit longer necks. Charles Darwin offered an alternative theory – evolution by natural selection. According to Darwin, individuals of the same species are all different from

TIMELINE

1801	1859	1866	1942
Jean-Baptiste Lamarck proposes the inheritance of acquired characteristics	Publication of Charles Darwin's *On the Origin of Species*	Gregor Mendel publishes work on inheritance in plants	Conrad Waddington coins the term 'epigenetics'

The beads of life

The nucleus of every cell in your body contains more than a metre (3ft) of DNA, packaged into 26 pairs of chromosomes. Each chromosome consists of one long double-stranded DNA molecule, which contains the information for synthesizing thousands of different cellular proteins. DNA is tightly coiled around barrel-shaped proteins called histones, producing a beads-on-a-string structure. This is coiled into helical fibres with a diameter of 30 nanometres (billionths of a metre), and these fibres are in turn organized into loops.

The combination of DNA and proteins inside a chromosome is referred to as chromatin. Chromosomes undergo continuous structural changes that regulate gene activity. Chromatin unzips at specific locations so that the molecular machinery that synthesizes proteins can gain access to the genetic information, then it closes up again once the gene is no longer active. These changes are controlled by epigenetic mechanisms, which 'mark' the chromosomes. Collectively, these marks constitute the epigenome.

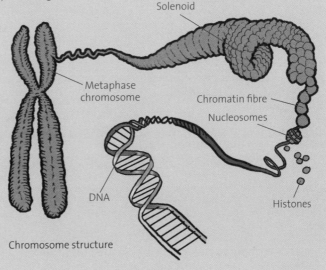

Solenoid

Metaphase chromosome

Chromatin fibre

Nucleosomes

DNA

Histones

Chromosome structure

one another. Some have beneficial variations, making them better adapted to their environment, while others have useless or harmful variations. Those with beneficial variations reproduce, while others are less successful and eventually become extinct.

1983
Timothy Bestor and Vernon Ingram identify DNA methyltransferase

2004
Michael Meaney et al. publish their landmark epigenetics study

2005
Rachel Yehuda et al. present research on the transgenerational transmission of trauma

Although Darwin did not fully realise it at the time, these variations occur in the form of genetic mutations. Subsequently, Gregor Mendel discovered the principles of heredity, and proposed that 'units of inheritance' (what we now call 'genes') are passed on to our offspring. As the evidence for natural selection accumulated, Lamarck's theory was refuted, then fell by the wayside. Epigenetics reconciles the two ideas by explaining how acquired characteristics can be inherited.

MOTHERLY LOVE

In 2004, Canadian researchers published a landmark study showing that the quality of maternal care has significant and long-lasting effects on offspring. They found that rat pups which are repeatedly licked and groomed by their mother during the first week of life were better at coping with stressful situations as adults than those who had received little or no contact with the mother.

The researchers then examined the animals' brains, and found that the differences in behaviour were due to alterations in the activity of the gene encoding the glucocorticoid receptor, which plays a crucial role in the stress response – those that had received high-quality care from their mothers in early life expressed far higher levels of the receptor in the hippocampus than those that had been neglected.

Transmitting trauma

Epigenetics is a form of 'genetic imprinting' – when cells divide, they pass on their epigenetic markers to daughter cells. And although neurons do not divide, epigenetic modifications to neuronal genes can be transmitted from one generation to the next. One poignant human example comes from a 2005 study showing that pregnant women who survived the attack on the Twin Towers transmitted their trauma to their children, by as-yet-unknown epigenetic mechanisms. Children born to women who were in the third trimester of pregnancy at the time of the attacks and who suffered post-traumatic stress disorder had lower cortisol levels later on in life, and also exhibited exaggerated distress responses.

In a follow-up study, the researchers examined the brains of suicide victims, some of whom had suffered childhood abuse, and people who had died suddenly from other causes. They found epigenetic differences in the brains of suicide victims who had been abused as children, altering activity of the glucocorticoid receptor gene and increasing their suicide risk.

EPIGENETICS IN ACTION

Epigenetics involves chemical modification of chromosomal DNA or associated proteins (*see page 157*). The best-known modification is methylation, in which a methyl group – a small chemical containing one carbon and three hydrogen atoms – is attached to a specific location in the DNA sequence or to a histone protein, by enzymes called methyltransferases. This 'marks' the chromosome so that its structure can be modified.

These 'mark ups' remodel the overall structure of chromosomes to affect the activity of genes. Epigenetic markers can have opposing effects on gene activity. Some open up a local area of the chromosome so that the genetic information can be used to synthesize proteins. Others close off the chromosome, effectively silencing the genes in that area.

Epigenetic mechanisms are implicated in virtually everything that the healthy brain does. They control the differentiation of neural stem cells during brain development, for instance, and contribute to the formation and maintenance of memories. During the ageing process, there is a change in the profile of epigenetic markers in the brain and elsewhere in the body, and neurological conditions such as Alzheimer's disease also involve epigenetic changes in the neurons. So, too, does drug addiction.

Epigenetic markers are, however, reversible. In the rat study, the Canadian researchers found that fostering neglected pups to caring mothers removed the epigenetic marks on the glucocorticoid receptor gene. They also found that the effects of maternal deprivation could be reversed if they gave the pups a chemical that inhibits the epigenetic modifications to the receptor gene caused by neglect. All of this has obvious and profound implications for child-rearing. Importantly, it suggests that all is not lost for children who were abused or raised in impoverished conditions.

> **THIS NEW AND EXCITING FIELD ... ALTERS THE WAY WE NEED TO THINK ABOUT OUR PAST AND OUR FUTURE.**
>
> Polish-born geneticist Eva Jablonka, 2010

The condensed idea
Life experiences can be inherited

40 Default mode

The wandering mind may provide fresh insights into brain function. Switching off from the outside world and entering the mental world of daydreams and imagination activates a set of brain regions called the default mode network, which is deactivated during demanding tasks. This network, which is disrupted in various diseases, may support the brain's core functions.

The brain is often thought of as an input/output system, processing information from the outside to generate an appropriate behavioural response, and most brain-scanning studies examine which areas become active, or 'light up', during a particular action or perception. It is a hungry organ that consumes about 20 per cent of the body's energy, despite accounting for just 2 per cent of its mass – yet we've known, since the 1950s, that its metabolic activity changes very little when it is actively engaged in performing a task.

In other words, the brain remains active when we are doing nothing at all, and has an intrinsic pattern of activity that uses up most of its energy. This 'baseline' activity is continuously running in the background, changing very little regardless of what we are doing. This is the brain's 'default mode', a network of brain regions that comes online when the brain is awake but resting. This activity may represent the brain's core functions, and it is affected by various neurological conditions. Many believe that studying the brain's resting state will reveal important insights into its function and dysfunction. And yet, the question remains: what exactly is the brain *doing* in its resting state?

TIMELINE

1929	1955	1974
Hans Berger proposes that the brain is always active, even at rest	Louis Sokoloff and colleagues discover that the brain's metabolic rate remains constant	David Ingvar collects data from resting-state brain scans and notes consistent activity patterns

ENGINE IDLING?

The brain's default mode was discovered entirely by accident, during brain-scanning experiments performed in the 1990s. Researchers noticed that participants' brains seemed to be active while they lay idle in the scanner, doing nothing in particular. At the time, researchers were still figuring out how to analyse brain-scan data properly, in particular how to separate signals associated with information processing from random signals produced by spontaneous activity. They initially dismissed the signals observed in resting state scans as noise, but then some began to wonder if they had characteristic properties of their own.

> [RESTING STATE ACTIVITY IS] ... THE "BRAIN WORK" WE CARRY OUT WHEN LEFT ALONE AND UNDISTURBED.
>
> David Ingvar, 1974

Until then, it had been taken for granted that the brain would generate spontaneous and unpredictable patterns of activity when not engaged in any particular task. But scans performed during rest periods showed the same low-frequency signal – specific regions of the brain consistently became active while participants lay quietly in the scanner with their eyes closed, doing nothing in particular. This set of brain regions came to be known as the default mode network (DMN); it is the best characterized of at least a dozen networks identified so far whose activity increases in the resting state.

The DMN comprises about half a dozen interconnected brain regions in the frontal and parietal lobe, which generate oscillating activity patterns with a frequency of 0.1 Hz (one cycle every ten seconds) or less. It includes the medial prefrontal cortex, which is involved in 'theory of mind', or making inferences about the intentions of others; the medial temporal lobe, which plays a critical role in memory; and the posterior cingulate gyrus, which, among other things, integrates activity of the frontal and temporal lobes. The DMN is negatively correlated with tasks involving paying attention to external stimuli. In other words, it becomes less active during tasks that involve focusing on the outside world, and more active when we turn our thoughts inwards.

2001	2007	2010	2011
Marcus Raichle and team report the discovery of the brain's default mode network (DMN)	Malia Mason and colleagues show that DMN activity is related to mind wandering	Discovery of the DMN in the rat brain	DMN discovered in the monkey brain

Resisting a rest?

The concept of the default mode network (DMN) plays a central role in neuroscience, but not everyone believes that it is of such great importance. Critics agree that understanding what the brain is doing during mind wandering will yield valuable insights, but argue that what we think of as the resting state may be something else altogether. They point out that the inside of a brain scanner is a claustrophobic and noisy environment, and that the brain activity observed when people are asked to do nothing may in fact correlate to a state of heightened vigilance – indicating that they are actively looking for something in the environment to which they can pay attention. And because the brain's metabolic activity changes very little, they say, it follows that resting-state activity is present to a greater or lesser extent during many different tasks, and therefore does not deserve to be treated as being a special type of activity.

THE WANDERING MIND

Consistent with the known functions of its subsystems, the DMN becomes active during internal modes of thoughts, such as remembering something that happened in the past, imagining a future event, or trying to put ourselves in someone else's shoes. This has led researchers to speculate that the main function of the default mode is to support mental activities collectively referred to as 'mind wandering': namely, daydreaming and imagination.

A key study published in 2007 provided some evidence that the default mode supports mind wandering. Researchers asked study participants to report how often their mind wandered during both an unfamiliar and a well-practised task – not surprisingly, they all said that their mind drifted off most during the task they knew well. Next the researchers scanned their brains while they performed both tasks, and found that the DMN was more active during the well-practised than the unfamilar task. The wandering mind, then, may provide important clues about what the brain evolved to do.

EFFECTS OF IMPAIRMENTS

Resting-state activity is disrupted in various neurological conditions. For example, one study showed that default mode network activity was reduced while autistic children performed a passive task and, conversely, that it was not deactivated during a cognitively demanding task – the opposite of what normally occurs in the non-autistic brain. It has been proposed that autism, characterized by impairments in social interaction, occurs because of an inability to simulate others' actions and intentions. This ability is linked to the medial prefrontal cortex, a part of the DMN, and atypical activity in this area is linked to the degree of social impairment, with those individuals who are most impaired showing the most atypical activity.

Other research shows that resting-state activity in two components of the DMN – the hippocampus and posterior cingulate gyrus – is reduced in patients with Alzheimer's disease compared to normal controls. During early stages of the disease, pathological plaques appear in key regions of the DMN. Some researchers therefore suggest that resting-state network activity could be used to detect and diagnose the condition. The DMN is also disrupted in patients with schizophrenia. One study showed that various symptoms of the condition, such as hallucinations, delusions and confused thoughts, are associated with increased activity in the medial prefrontal cortex and posterior cingulate gyrus during rest.

The condensed idea
The brain's 'dark energy' may be at the heart of its core functions

41 Brain-wave oscillations

Large groups of nerve cells exhibit rhythmic patterns of activity called brain-wave oscillations, which differ in size, frequency and timing. The synchronization of these patterns within and between different areas of the brain is thought to play a key role in the processing of neural information.

Individual neurons exhibit spontaneous, rhythmic patterns of electrical activity in the form of action potentials (*See also Chapter 4: The nervous impulse*), and cells located in the same part of the brain form nuclei, or clusters, that send their axons to the same target brain region. Such organization is laid down during brain development to form functional networks with extremely precise neural pathways. Cells within these pathways are connected to each other not only by chemical synapses, but also by electrical synapses, or gap junctions, which enable them to coordinate their activity and fire simultaneously.

The synchronized bursts of electrical activity generated by large neuronal populations containing millions of cells are known as brain waves, and can be detected from outside the skull using two brain-imaging techniques called electroencephalography (EEG) and magnetoencephalography (MEG). Ever since their discovery in the late 1920s (*see page 166*), we've known that brain waves have distinct patterns, but only in recent years have we begun to appreciate the significance of these patterns for information processing.

TIMELINE

1875	1924	1964
Richard Caton observes spontaneous electrical rhythms in the brains of mammals	Hans Berger invents the electroencephalogram to record brain waves	Early report on gamma waves based on electrode recordings from the visual cortex of monkeys

RHYTHMS OF LIFE

Brain-wave rhythms are usually characterized by three properties. One property is frequency, measured in Hertz (Hz, or cycles per second); brain-wave frequencies vary from ultraslow oscillations with a frequency of less than 1 Hz to ultrafast oscillations with a frequency greater than 600 Hz. Another property is the amplitude, or size, of the wave; when recorded using EEG, brain-wave amplitude typically varies between 1 and 10 microvolts, or thousandths of a volt. The third property is the phase, or timing, of the wave; this can be altered to synchronize the activity of neurons within and between different brain regions, a process known as 'phase-locking'.

There are at least a dozen different patterns of brain-wave oscillations, including:

Alpha waves: low-amplitude oscillations with a frequency of between 8 and 12 Hz. They are associated with a calm, relaxed state and are generated primarily by the occipital lobe.

Beta waves: low-amplitude oscillations with a frequency of between 12 and 30 Hz. They are linked to the alert state of normal, waking consciousness, and are produced in the frontal lobes when we concentrate and generate voluntary movements.

Gamma waves: high-amplitude oscillations with a frequency of between 20 and 100 Hz. Gamma waves triggered by the occipital lobe have been associated with attention, and it has been suggested that those in the 40 Hz frequency band play an important role in consciousness.

Theta waves: low-amplitude oscillations with a frequency of between 4 and 7 Hz. They are particularly strong in the hippocampus, and are connected to learning and memory.

1990	**1998**	**2012**
Francis Crick and Christof Koch suggest that gamma waves are critical for consciousness	First experimental evidence linking gamma waves to visual binding	Andreas Engel and colleagues use MEG to examine oscillation frequency in different parts of the brain

'Psychic energy'

Electroencephalography (EEG) was invented in the 1920s by the German psychiatrist Hans Berger. Berger was interested in psychic phenomena such as telepathy, and believed that they had an underlying physical basis – a 'psychic energy' – that could be transmitted between individuals. Berger was initially interested in how the brain's blood flow changes in relation to mental activity. Then, taking inspiration from the work of Richard Caton, a British physiologist who had discovered spontaneous electrical activity in the brains of rabbits and monkeys during the 1870s, he began using electrodes to measure the brain's electrical activity in patients who had had holes cut out of their skulls before surgery. Berger recorded the first human EEG in 1925, and quickly noticed that there were different patterns of brain waves. He was the first to describe alpha waves – these are, therefore, also known as Berger waves. He noted that alpha waves were suppressed, or substituted, by higher frequency beta waves when his subjects opened their eyes, indicating a transition from a relaxed to a focused state.

First EEG recording, drawn by Hans Berger

Different parts of the brain generate brain-wave oscillations in different frequency bands to avoid signal jamming and perhaps to enable multiple, overlapping frequencies to be used for communication. For example, medial temporal lobe structures such as the hippocampus produce oscillations predominantly in the theta frequency range (4–6 Hz); regions on the outer surface of the parietal lobe operate in the beta frequency range (12–30 Hz); and sensory and motor areas generate oscillations of even higher frequencies (32–45 Hz). This may be linked to different types of neurons or to regional variations in cellular organization. Similarly, cells within discrete layers of the cerebral cortex generate oscillations of different frequencies.

BEYOND SLEEP RESEARCH

Brain-wave oscillations were traditionally studied within the context of sleep, which consists of several distinct stages, each characterized by a distinct brain-wave pattern. In the past few decades, however, researchers have come to realize that brain-wave rhythms play important roles in numerous mental activities. We are only just beginning to understand the role that brain-wave oscillations play in mental functions and behaviour, but our knowledge of these processes will improve as brain-imaging techniques become more sophisticated.

When activated during some task, individual cells can 'reset' the timing of their electrical activity to synchronize it with the frequency of the brain-wave oscillations of the surrounding tissue. This is thought to facilitate information processing. The synchronized oscillations within populations of neurons unify them into a functional network, and synchrony between neuronal populations in distant parts of the brain may coordinate their activities and facilitate the transfer of information between them. Synchrony can occur when oscillations of the same frequency are coupled, or 'locked,' together, or when low-frequency oscillations are 'nested' within higher frequency ones.

> **SYNCHRONIZATION OF NEURONAL DISCHARGES CAN SERVE FOR THE INTEGRATION OF DISTRIBUTED NEURONS ... [AND] MAY UNDERLIE THE SELECTION OF BEHAVIOURALLY RELEVANT INFORMATION.**
> Andreas Engel, 1999

When rats perform spatial navigation tasks such as finding their way through a maze, oscillatory rhythms in the hippocampus are dominated by theta waves, and this pattern is thought to enhance the encoding of spatial memories. Theta oscillations are also associated with hippocampal 'replay', during which memory traces are reactivated after encoding so that they can be consolidated, or strengthened.

Gamma waves play a role in a phenomenon of consciousness known as the binding problem, whereby action potentials in different brain regions produce a unified perception of a single object. Seeing a red square and a blue circle simultaneously, for example, generates impulses that all look the same, but how does the brain 'know' that red belongs to the square, and blue to the circle? An early study published in 1988 showed that visual stimuli cause neurons in the visual cortex of monkeys to oscillate synchronously with a frequency of 40 Hz, leading prominent researchers to suggest that this frequency band is critical for visual awareness and consciousness.

The condensed idea
Patterns of brain-wave rhythms contribute to information processing

42 Prediction error

The brain is constantly dealing with uncertainty. It learns models of the outside world based on limited – and often ambiguous – information, and uses pre-existing knowledge to make predictions about the causes of sensory experience. Our perceptions are the brain's 'best guess', as it minimizes the discrepancy between its predictions and actual outcomes to refine its world model.

E very waking second, the brain processes huge amounts of information that it receives from the body and the outside world. Much of this information is noisy, or ambiguous and uncertain, and yet the brain uses it to learn an accurate internal model of the self and the world around us, which it then uses to guide our behaviours. To do so, it generates inferences, or predictions, and compares them to the actual sensory inputs it receives, in order to determine the causes of bodily sensations. These inferences constitute beliefs about the world, which are updated when a mismatch occurs between expectations and outcome.

The term 'prediction error' refers to such a discrepancy between the brain's predictions and the incoming sensory information. According to this concept, the brain learns from its mistakes, using errors in its predictions to minimize unexplained sensory fluctuations and refine its internal model of the world – the smaller the error, the more accurate the brain's world model. Neuroscientists now use the concept as a unifying theory of brain function,

TIMELINE

1812	1866
Pierre-Simon Laplace publishes his formulation of Bayes' theorem	Hermann von Helmholtz describes visual perception as unconscious inferences based on knowledge

The Bayesian brain

Thomas Bayes was an 18th-century mathematician who proposed a statistical theorem to express how beliefs should change in the light of new evidence. This process, known as Bayesian inference, regards the degree of a belief in probabilistic terms. Bayes' theorem explains how relevant new information changes the probability that a given belief is correct. It is most commonly expressed as an equation that determines the relationship between the probabilities (P) of two events (A and B). Many computational neuroscientists now regard the brain as a Bayesian probability machine that makes inferences about the external world then updates them according to sensory information. As such, the brain treats ambiguous sensory information statistically, in terms of the probability that a given prediction will turn out to be correct. As new information becomes available, it changes the probability that a given prediction is correct, and alters its internal models accordingly.

$$P(A|B) = \frac{P(B|A) \ P(A)}{P(B)}$$

Equation for Bayes' theorem

to explain how perception, cognition and actions work together to minimize these errors, and to illustrate how thought processes go awry in psychiatric disorders such as schizophrenia.

FILLING IN THE GAPS

Visual perception is perhaps the best example of how the brain makes inferences based on limited and ambiguous information. Perception is often regarded as a 'bottom-up' process, in which sensory inputs entering the eyes are processed in hierarchical and increasingly complex stages, to reconstruct the visual scene that we experience. In fact, as the pioneering

1983

Geoffrey Hinton suggests that the brain makes decisions based on uncertainty in the outside world

1980s

Chris Frith proposes that prediction error failures can explain schizophrenic symptoms

2005

Karl Friston proposes the free energy principle

experimental psychologist Hermann von Helmholtz suggested in the 19th century, perception also involves 'top-down' mechanisms – it requires a great deal of inference, whereby the brain uses prior knowledge to make sense of incomplete visual information.

When we look at the world around us, we rarely see objects in their entirety. Often, we see only parts of them, because of the angle from which we are looking at them, and because some sections may be obscured by other objects. Nevertheless, our beliefs about visual perceptions are accurate enough that we can usually identify objects correctly. We can, for example, easily identify a chair, even if it is partly hidden under a table, and even if we have never seen that particular type of chair in the past. This is because the brain fills in the gaps that are missing from the sensory information, and uses pre-existing knowledge of the form and function of this class of objects to make an inference about what it is that is being perceived. In other words, although the sensory information may be incomplete, it is often more than enough to confirm our expectations.

The free energy principle

Free energy has been proposed as a measure of the amount of unpredictable fluctuation in self-organizing systems. In the nervous system, it equates to the degree of discrepancy between expectation and outcome. The free energy principle states that the brain acts to minimize free energy to a minimum, in order to encode information as efficiently as possible and to ensure that exchanges with the world are predictable.

ACTION AND AGENCY, SELF AND OTHER

Generating voluntary movements is one of the primary functions of the brain. As we plan a movement, the brain generates a forward model that predicts the sensory and behavioural consequences of that action. It then compares its prediction with the actual outcome. A close match between prediction and outcome is essential for agency – the sense that we are in control of our actions – which in turn is a critical component of self-consciousness. We also simulate other peoples' actions, in order to make inferences about their behaviour and intentions, an ability known as theory of mind.

A recent brain-scanning study showed that prediction errors for our own and other people's decisions are encoded in different regions of the prefrontal cortex. Researchers scanned participants' brains while they performed a simple value-based decision-making task and while they simulated others' behaviour to predict how they would perform on the same task. This involved predicting not only how the other people will act, but also what reward they will receive at the end. The larger the prediction error in the simulated action, the more activity observed in the dorsolateral and dorsomedial prefrontal cortex. Similarly, the larger the prediction error in the simulated reward, the more active the ventromedial prefrontal cortex.

CODING ERRORS

People with schizophrenia may experience auditory and visual hallucinations, whereby internally generated stimuli are incorrectly perceived to take place in the outside world. They may also have a disrupted sense of agency, which causes them to have delusions of control – misattributing their thoughts and actions to external forces. All of these symptoms can be explained in terms of how the brain encodes and responds to prediction errors.

A study published in 2010 shows that patients with schizophrenia have difficulty making accurate predictions about the sensory consequences of their actions, and that the less precise their predictions, the more severe their delusions of control. Delusions and hallucinations are ambiguous experiences that patients explain with implausible beliefs. A schizophrenic patient, for example, may hear voices in their head and mistakenly attribute them to evil forces. This may occur because patients are too reliant on external stimuli, and do not evaluate their accuracy. Their internal models are not updated as they should be by their sensory experiences. Consequently, their internal models of the world are not consistent with reality, and their implausible beliefs are maintained.

THE WHOLE FUNCTION OF THE BRAIN IS SUMMED UP IN: ERROR CORRECTION.

British psychiatrist
Ross Ashby, 1954

The condensed idea
The brain is an inference machine

43 Neural stem cells

The brain contains several areas of self-renewing stem cells, raising the possibility that it can repair itself following an injury. Neural stem cells can also be used to grow brain cells in the lab, offering hope of cell-transplantation therapies for various neurological conditions. The full potential of neural stem cells is yet to be realized, however.

Stem cells are immature, non-specialized cells that can differentiate into specialized cell types, and are characterized by several properties. They are self-renewing – in other words, they can divide to make copies of themselves. They are also pluripotent – that is, they have the ability to differentiate into any of the specialized cell types found in the body. Neural stem cells are more specialized yet: they are limited to producing all of the different types of neurons and glia found in the brain, and are said to be multipotent.

Embryonic stem cells, as their name suggests, are found in the embryo. During the earliest stages of development, the human embryo consists of a sphere of stem cells that, although initially identical, go on to form every type of cell in the body. The cells used for scientific research are derived from artificially fertilized eggs that are then grown in Petri dishes.

Embryonic stem cell research is extremely controversial, and has proven to be a political hot potato in many countries. In adults, stem cells are found

TIMELINE

1981	1992	1998
Embryonic stem cells first isolated by researchers in Cambridge and San Francisco	Brent Reynolds and Sam Weiss isolate neural stem cells from the adult mouse brain	Fred Gage and colleagues identify neural stem cells in the adult human brain

in most organs of the body, where they are essential for maintaining and repairing tissues. Bone marrow, for example, contains haematopoietic stem cells, which form the different types of blood cells, while skin contains other stem cells that replace the dead cells which are constantly being sloughed off the skin's surface.

Until recently, the brain was thought to be an exception. In the 1990s, however, researchers discovered neural stem cells in the brains of mice, and then in the human brain. Neurons generated from stem cells in the adult mouse play important roles in brain function, but it's still not clear whether this is true of the human brain, too. There is also some evidence that neural stem cells may contribute to the formation of brain tumours. Nevertheless, neural stem cells

Lab-grown brain parts

Japanese researchers have found that embryonic stem cells from mice can organize themselves into complex three-dimensional structures when grown in a floating suspension and fed with the right combination of signalling molecules. In early experiments, they used this specially developed technique to grow functional brain tissue, which they successfully transplanted into the brains of newborn mice. More recently, they have used stem cells to grow parts of the pituitary gland, which are fully functional when transplanted into mice. In their most recent experiments, the researchers used human embryonic stem cells to grow retinal tissue, complete with light-sensitive photoreceptors. The work is at the cutting edge of an emerging area of research called tissue engineering, and could lead to new therapies for neurological diseases. Eventually, researchers will be able to grow brain tissue containing specific types of neurons, which could be grafted into the brain, and lab-grown retinal tissue could one day be used to restore sight to people with macular degeneration.

2001	2006	2009	2012
USA bans federal funding for new stem-cell research	Induced pluripotent stem cells produced from mouse muscle cells	USA lifts the ban on federal funding for stem-cell research	Yoshiki Sasai and colleagues grow retinal tissue from human embryonic stem cells

obtained from the adult human brain can be grown in the lab and used to generate mature, functional brain cells of all types: neurons, astrocytes and oligodendrocytes (*see pages 12–13*). This offers hope that they could be used to develop treatments for neurological diseases, and numerous research groups around the world are now pursuing this goal.

OVERCOMING CHALLENGES

Neural stem cells could be used to develop treatments for a wide variety of neurological conditions, including Alzheimer's, Parkinson's and motor neuron diseases, spinal cord injury and stroke. To this end, researchers are exploring two approaches. One is to coax into action the neural stem cells already present in the brain. The other is to use embryonic or neural stem cells to grow specific types of mature neurons in the lab, then transplant them into the brain.

Of these two strategies, the second is more promising. Decades of research into the mechanisms of brain development have revealed much detail about how embryonic stem cells are specified to generate different kinds of mature neurons and glial cells. With this knowledge, researchers can now use human embryonic stem cells to grow, for example, the dopamine-producing midbrain neurons that die off in Parkinson's disease, or the motor neurons that die in amyotrophic lateral sclerosis and related conditions. What's more, neurons produced in this way can alleviate disease symptoms when grafted into the brains of animals.

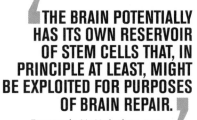

THE BRAIN POTENTIALLY HAS ITS OWN RESERVOIR OF STEM CELLS THAT, IN PRINCIPLE AT LEAST, MIGHT BE EXPLOITED FOR PURPOSES OF BRAIN REPAIR.

Fernando Nottebohm, 2011

Translating the animal studies into therapies was initially challenging. Researchers faced a number of technical difficulties, such as how to target cells to the appropriate part of the brain, and how to keep the implanted cells alive for sufficient periods of time. These challenges have now been overcome, and human clinical trials are underway for stroke, spinal cord injury, amyotrophic lateral sclerosis and Parkinson's disease. The trials involve injecting stem cells into the affected region, in order to replace the cells that have been lost due to injury or disease. Such therapies are still in the earliest stages, but the initial results seem promising, and the tremendous rate of discovery in stem-cell research in recent years will no doubt accelerate the development of treatments.

CELLS OF HOPE

Cells taken from skin, muscle and other parts of the adult human body can be reverted into stem cells resembling those found in the embryo. These so-called 'induced pluripotent stem cells' can then be reprogrammed to differentiate into various kinds of brain cells or other specialized cells. This, too, is based on developmental research – the procedure of reprogramming involves introducing specific genes into the cells, which drive them to differentiate along a certain pathway, or to de-differentiate into an embryo-like state.

The results are certainly exciting. In 2008, American researchers took skin cells from an 82-year-old woman with amyotrophic lateral sclerosis, reverted them to pluripotent stem cells, then converted those to motor neurons. And in 2011, a team of Japanese researchers showed that the intermediate step in this process can be skipped. They obtained fibroblasts (connective tissue cells) from patients with Alzheimer's disease and converted them directly into functioning neurons.

Induced pluripotent stem cells offer a promising new approach to studying neurological disease. Cells from patients with neurological disorders could be converted into neurons and grown in the lab so that researchers can examine the cellular mechanisms of disease. Several recent studies suggest that induced pluripotent stem cells may have genetic abnormalities, however, and this raises doubts about how useful they might be.

The condensed idea
Stem cells could repair the damaged brain

44 Brain stimulation

Electrical and magnetic stimulation can be used to alter brain activity. These techniques can also be used to examine the brains of patients about to undergo neurosurgery, to investigate brain function in the lab, and to facilitate rehabilitation following brain injury – as well as to enhance cognitive function in healthy people.

Electrical brain stimulation conjures up images of barbaric electroshock treatments meted out to psychiatric patients, as depicted in Ken Kesey's book *One Flew Over the Cuckoo's Nest*. Certainly, electroconvulsive therapy was used widely in the past, but it is rarely used today. Other methods of electrical stimulation are useful in the clinic and the laboratory, however, and some could prove to be effective treatments for neurological and psychiatric conditions.

A BRIEF HISTORY

The history of electrical brain stimulation began in the 1870s, with the experiments of Eduard Hitzig, a German doctor who worked at a military hospital during the Franco-Prussian war. Hitzig treated many soldiers with head wounds – some of whom had had bits of their skulls blown away – using wires attached to a battery in order to apply electrical currents to their exposed brains. Later, with Gustav Fritsch, he strapped dogs to his wife's dressing table and applied electrical currents to their brains. This showed that stimulation of a certain part of the brain (the motor cortex) elicited movements on the opposite side of the body.

TIMELINE

1771	1870	1920s
Luigi Galvani discovers bioelectricity	Eduard Hitzig and Gustav Fritsch publish their work on electrical stimulation of the dog brain	Wilder Penfield pioneers electrical stimulation of the brain in conscious neurosurgical patients

In the 1920s, the neurosurgeon Wilder Penfield pioneered the use of electrical stimulation to examine the brains of conscious epileptic patients before operating on them. Surgery, a last resort for patients who don't respond to drugs, requires careful assessment of the brain tissue producing the seizures, so that surrounding tissues which might perform important functions aren't damaged during the procedure.

> **PASSING CURRENTS THROUGH THE HEAD, ONE COULD EASILY PRODUCE MOVEMENTS OF THE EYES.**
> Hitzig and Fritsch, 1870

Penfield administered local anaesthetic to these patients, opened up their skulls, and stimulated the brain in and around the area causing their seizures. Because they were conscious throughout the evaluation, the patients could report on the effects of the stimulation. Thus, Penfield delineated the abnormal tissue and spared adjacent tissues involved in functions such as speech, memory and movement. His procedure also enabled him to map the brain regions involved in sensory and motor functions, the results of which are still widely used today.

A 'PACEMAKER' FOR THE BRAIN

Deep brain stimulation (DBS) is an experimental procedure involving the implantation of thin wire electrodes into a specific part of the brain. The implanted device is attached to a battery that is placed beneath the skin on the chest or attached to the inside of the skull. The device, often referred to as a 'brain pacemaker', alters activity in the target brain area by delivering regular electrical impulses to it.

In 2002, the US Food and Drug Administration approved DBS as a treatment for Parkinson's disease, and since then it has been given to approximately 80,000 patients with the condition. The National Health Service also offers the treatment to small numbers of patients. When targeted to one of several regions involved in movement control, DBS can alleviate the symptoms. The

1938	1985	2002
Ugo Cerletti and Lucio Bini introduce electroconvulsive therapy as a psychiatric treatment	Anthony Barker and colleagues publish the first study of the effects of TMS on human beings	FDA approves DBS as a treatment for Parkinson's disease in the USA

DIY brain stimulation

The do-it-yourself biology movement has gained momentum in recent years, with some people applying tDCS to their own brains in the hope that it might improve their mental abilities. The device can be built very easily, using cheap, off-the-shelf electrical components – all that's needed is a 9-volt battery and some electrode cables – and there is now at least one American company that sells tDCS kits. It touts tDCS as 'one of the coolest health/self-improvement technologies there is', and sells its kit to wannabe brain hackers for $99. When used under properly controlled, experimental conditions, tDCS is considered to be safe and to have few, if any, harmful side effects. In experimental conditions strict limits are placed on the size and duration of the current and the frequency of stimulation. People experimenting on themselves, however, may be putting themselves at risk if they don't adhere to these guidelines.

technique is now also being used as an experimental treatment for other conditions, including depression, obsessive-compulsive disorder and addiction.

In 2012, neurosurgeons operating on epileptic patients reported that DBS can enhance the formation of spatial memories when targeted to a part of the brain called the entorhinal cortex. Although preliminary, these findings suggest that DBS could effectively improve memory deficits in patients with Alzheimer's disease, and could also be used to enhance memory function in healthy individuals.

Despite its widespread use, it's still not clear exactly how DBS works, or why it is more beneficial for some patients than for others. The technique involves brain surgery and therefore carries the risk of infection and permanent brain damage. What's more, it often produces unwanted side effects, and the long-term effects are still unclear.

IF THE CAP FITS …

Transcranial direct current stimulation (tDCS) is a non-invasive electrical stimulation technique, in which a constant electrical current is applied to the surface of the brain by a tight cap containing electrodes. This alters the electrical properties of neurons in the targeted area, leading to increased or decreased activity, depending on how the current is applied.

Exactly how tDCS works is unclear, but early research suggests that it can help stroke patients relearn how to move and speak, and clinical trials suggest that it could benefit people with Parkinson's disease. Preliminary evidence also suggests that it can enhance processes such as attention, memory and motor-skill learning in healthy people. Based on these findings, the US military now uses tDCS to train its snipers and pilots.

MAGNETIC EFFECTS

Transcranial magnetic stimulation (TMS) is another non-invasive brain stimulation technique that uses a figure-of-eight-shaped coil to deliver small, rapidly changing magnetic fields to specified parts of the brain. This induces weak electrical currents in the brain, which enhance or disrupt the activity of neurons.

TMS is used in the clinic for both diagnostic and therapeutic purposes. It can be employed to examine the function of specific parts of the brain, for example, and to evaluate the damage caused by stroke, motor neuron disease and multiple sclerosis. The technique has also been tested as a treatment for a variety of neurological and psychiatric conditions, but most studies show that it has only modest effects. It may, however, be of therapeutic value to some patients with major depressive disorder.

TMS is also widely used to examine brain function in the laboratory. A typical experiment involves interfering with the activity of a specific, targeted brain region, to determine whether or not it is involved in a given task or process. Participants may, for example, be asked to write something, while the researcher zaps their motor cortex to interrupt neural control of the hand.

The condensed idea
Zap caps could help repair and improve the brain

45 Cognitive enhancement

The term 'cognitive enhancement' refers to a variety of methods that boost brain function, but it is used most often in the context of 'smart drugs', which aim to improve mental abilities such as attention and memory. The use of smart drugs has increased in recent years – this has raised concerns over safety, and has ethical implications for society.

Traditionally, people have improved their brains through education and various forms of mental training, and most cultures have used caffeine and nicotine to increase alertness for centuries. Modern neuroscience now offers new ways to enhance cognitive function, collectively referred to as cognitive enhancement – these include invasive and non-invasive brain stimulation, as well as the use of so-called 'smart drugs'.

HACKING THE BRAIN
The brain can be stimulated non-invasively, using transcranial direct current stimulation (tDCS) and transcranial magnetic stimulation (TMS). Both techniques are used experimentally and clinically, but there is some evidence that they can boost cognitive function in healthy people. A growing band of DIY 'brain hackers' *(see page 178)* apply these methods on themselves – although it's not clear how safe or effective they are when used outside of properly controlled laboratory conditions.

TIMELINE

1917	1948	1964
Karl Lashley discovers that strychnine facilitates learning in rats	CIBA Pharmaceuticals (now Novartis) trademarks the brand name Ritalin	Corneliu Giurgea develops Piracitam, the first smart drug

Deep brain stimulation (DBS) is an experimental surgical technique in which electrodes are implanted into specified regions of the brain. It is used as a treatment for Parkinson's disease and various other conditions, such as depression and obsessive-compulsive disorder, but preliminary results suggest that it could also enhance mental functions such as memory in healthy individuals. DBS is highly invasive, so its use for the purposes of cognitive enhancement is unlikely to become widespread.

'SMART' DRUGS?

The most common method of cognitive enhancement is the use of smart drugs, or 'cogniceuticals'. Many are prescription drugs for specific neuropsychiatric or medical conditions, but their use among healthy people is rising. Others are designed and marketed specifically for use in cognitive enhancement. The term 'smart drug' is something of a misnomer, because these substances don't actually make people more intelligent. Instead, they can make healthy people more productive, by targeting and enhancing processes such as attention, concentration and memory. Many different types of smart drugs exist, the most widely used being psycho-stimulants normally prescribed as medication, including:

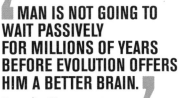

MAN IS NOT GOING TO WAIT PASSIVELY FOR MILLIONS OF YEARS BEFORE EVOLUTION OFFERS HIM A BETTER BRAIN.

Corneliu Giurgea, 1970s

Dextroamphetamine (marketed as Adderall in the United States): a mixture of amphetamine salts used in the treatment of attention deficit hyperactivity disorder (ADHD). It acts mainly on the dopamine system, and increases the concentration of this neurotransmitter at synapses by two mechanisms: it prevents neurons from mopping up dopamine after it has been released, by binding to a cell surface protein called the dopamine transporter, and it also inhibits an enzyme called monoamine oxidase, which normally breaks down dopamine,

1983	1986	2002	2012
Research and development on donepezil begins	Modafinil offered as a treatment for narcolepsy in France	Stanford researchers discover that donepezil improves memory	Itzhak Fried and colleagues report that DBS enhances spatial memory function

Memory-boosting drugs

Nearly a century ago, the psychologist Karl Lashley discovered that strychnine enhances learning in rats. Strychnine is best known as a poison and is used as a pesticide, but at low doses it acts as a stimulant by inhibiting the enzyme acetylcholinesterase, which breaks down the neurotransmitter acetylcholine. Donepezil, an Alzheimer's drug developed in the 1980s, inhibits the same enzyme. It slows memory loss in Alzheimer's patients and also improves cognitive function in healthy people, so it could be used as a smart drug, too. Decades of animal research have revealed other key molecules involved in memory. In 2000, Eric Kandel of Columbia University won a Nobel Prize for his work on sea slugs, which showed that learning occurs by several different synaptic mechanisms, and requires an enzyme called CREB. In 1998, Kandel launched a pharmaceutical company to develop drugs that increase CREB levels in the brain, which will be marketed for Alzheimer's and age-related cognitive decline in healthy people.

as well as serotonin, adrenaline and noradrenaline. The actions of Adderall on the prefrontal cortex improve working memory function and the ability to focus.

Methylphenidate (Ritalin): is another stimulant used to treat ADHD. Its molecular structure resembles that of amphetamine, and it works in a similar way – by exerting effects on the dopamine neurotransmitter system in the prefrontal cortex. First approved by the US Food and Drug Administration as a treatment for ADHD in 1955, Ritalin was prescribed widely for treatment of the condition during the 1990s. It is also licensed and sold as a treatment for ADHD in Canada, Australia, and several European countries, including the U.K.

Modafinil (Provigil): a non-amphetamine stimulant prescribed to treat fatigue associated with narcolepsy. Because modafinil can keep people awake and alert for long periods of time, it is also prescribed to those working long hours and nightshifts. Modafinil is now used by the military forces of the USA, Britain, China, France and India to combat the effects of sleep deprivation, which has severe, detrimental effects on troops' performance.

Smart drugs are now readily available on the Internet, and their non-medical use is increasing. The trend seems to be particularly strong within universities. According to a survey conducted in 2005, nearly 7 per cent of students in US universities have used smart drugs, and in some campuses that figure reached 25 per cent. In a reader survey in the scientific journal *Nature* several years later, one in five of approximately 1,400 people who responded – mostly academics and other professionals – admitted to having used a prescription drug to improve their attention or focus. Of those who responded, Ritalin was the drug most frequently used. The trend also seems to be growing among teenagers, who are faced with increasing pressure to improve their performance at school.

TO TAKE OR NOT TO TAKE?

The issue of smart drugs has divided both the scientific community and the general public. Some people openly advocate their use, arguing that safe usage can be of great benefit to individuals and society. Others disagree. They regard using smart drugs as 'cheating', and contend that it could create problems for society. In sports, the use of performance-enhancing drugs is tightly regulated. Should there be similar regulations for cognitive enhancement in schools and the workplace? Would widespread usage place pressure on those who do not wish to use them to conform? And how about those who want to use them but cannot afford to do so? Cognitive enhancement could also increase social inequalities.

IF WE ARE SMART ENOUGH TO INVENT TECHNOLOGY THAT INCREASES BRAIN CAPACITY, WE SHOULD USE THAT ADVANTAGE.

Michael Gazzaniga, 2010

Safety and efficacy are other major factors. Neurotransmitter systems are extremely complex, and manipulating them could have undesired – and unknown – consequences. Even if smart drugs are safe in the short term, the effects of long-term use by healthy people still have not been examined. What's more, we don't really know how effective they are. Laboratory tests examining their effects on performance in various cognitive tasks have provided mixed results. Almost certainly, the effects of smart drugs will differ between people, and perhaps in the same person depending on the situation.

The condensed idea
Drugs and technology can boost brain function

46 Brain scanning

The terms 'brain scanning' and 'neuroimaging' usually refer to functional magnetic resonance imaging (fMRI), a method that measures brain activity indirectly and is often used to show which parts of the brain 'light up' during mental tasks. We still don't know exactly how this works, and the methods for interpreting brain-scanning data have often been criticized.

The word 'brain scanning' covers a variety of techniques that can be used to visualize the structure of the living human brain and how it responds to various stimuli.

Electroencephalography (EEG) involves recording the electrical activity of large populations of neurons near the surface of the cerebral cortex, using electrodes placed onto the scalp.

Magnetoencephalography (MEG) is similar to EEG, but picks up the magnetic fields produced by the brain's electrical activity. It's more sensitive than EEG, but also more expensive.

Positron emission tomography (PET) is a technique that tracks the movements of radioactive 'tracer' compounds injected into the body. It can be utilized to measure blood flow, energy use and the levels of receptor proteins and other important molecules.

TIMELINE

1890	1929	1961
Charles Roy and Charles Sherrington link blood flow to brain cell metabolism	Hans Berger invents the EEG	James Robertson and colleagues build the first PET scanner

Magnetic resonance imaging (MRI) is also employed to obtain images showing brain structure, and can be used to detect the structural changes that occur in conditions such as Alzheimer's disease. It has largely replaced X-rays because it's safer and gives more detailed images.

Diffusion tensor imaging (DTI) is a variation of MRI that measures signals produced by the movement of water molecules in the brain. It enables experts to visualize the brain's white matter tracts, which contain large bundles of myelinated axons that connect distant brain regions to each other.

Functional magnetic resonance imaging (fMRI) uses changes in blood oxygen levels to measure brain activity. It has advantages over EEG and PET, because it can be used to show activity in the deepest parts of the brain, yet it is non-invasive because it does not involve injecting radioactive substances.

The term 'brain scanning' usually refers to fMRI, the use of which has increased significantly in the past ten years, and which has captured the public imagination because of the beautiful brain images it can produce. Thousands of fMRI studies are published every year, many of them reported by the mass media with simplistic headlines along the lines of: 'Scientists find the brain area responsible for X.' In recent years, however, neuroscientists have realized that fMRI results are far more complex.

THE BRAIN AT WORK

fMRI uses powerful magnets to detect changes in blood oxygen around the brain. The method is based on the assumption that active brain cells need extra energy in the form of oxygen. The signals fMRI detects are referred to as the blood-oxygen-level-dependent (BOLD) response, and are an indirect measurement of brain activity.

1968	1990	2009
David Cohen measures MEG signals for the first time	Seiji Ogawa and colleagues develop fMRI	Yevgeniy Sirotin and Aniruddha Das discover pre-emptive blood flow

A typical fMRI study involves scanning peoples' brains while they perform some mental task, then comparing the data to a baseline period when they are not doing anything in particular. This produces a three-dimensional image showing how activity in different brain regions is related to the task. fMRI scans often produce tens of thousands of individual data points, which are referred to as voxels (or volume pixels). Each voxel corresponds to a tiny cube of brain tissue of approximately one cubic millimetre, containing about 50,000 neurons. Researchers often focus on the voxels in a small number of predetermined 'regions of interest', and compare their activity during the task and at rest to make inferences about the relative changes.

BOLD CRITICISMS

One problem with fMRI is that the method measures brain activity indirectly. We still do not know how BOLD signals are produced by the brain, or exactly how they are related to neuronal activity, and the assumption that increased blood flow to a region of the brain means increased activity in that region has been called into question (*see box*).

The BOLD signals detected by fMRI are also small and 'noisy', meaning that complex statistical procedures are needed to detect the brain activations.

One step ahead

fMRI is based on the assumption that enhanced brain activity is related to increased blood flow, but it cannot be taken for granted that this is always the case. In 2009, researchers from Columbia University scanned monkeys' brains while they viewed pictures and found that activity in the visual cortex closely corresponded to increased blood flow to that region, as expected. To their surprise, however, they saw the same increases in blood flow even when the monkeys were not shown a visual stimulus. They concluded that the brain anticipates which regions will become active in the near future, and pre-empts this by increasing the blood flow to those regions, even though they do not necessarily become active. These findings therefore suggest that increased blood flow is not always associated with higher levels of neuronal activity, and raise doubts about the validity of fMRI data.

When used properly, these statistics are valid, but some researchers do not follow the appropriate controls. What's more, we now know that factors such as breathing and head movements can contribute to the BOLD signal, and a recent study showed that the same fMRI data can yield different results when analysed with different software packages.

The strongest criticisms of the technique are levelled at how the data are interpreted. Researchers often make what are referred to as 'reverse inferences' when interpreting fMRI data. Imagine a hypothetical fMRI study showing that brain area A is activated during task X. The researchers look at earlier studies and find that the same area is also active during mental process Y. They therefore conclude that task X relies on mental process Y. Unfortunately, although it seems attractive, the logic behind this reasoning is flawed – even if area A consistently lights up during task X, we cannot conclude that task X is always being performed when we see activity in area A.

> THE MORE YOU LOOK [AT FMRI SCANS], THE MORE YOU GET MEANINGFUL INFORMATION. WHAT PREVIOUSLY WAS NOISE IS NOW SUDDENLY SIGNAL.
>
> Neuroscientist Peter Bandettini, 2012

Reverse inference is related to the misleading and simplistic idea that a single part of the brain is responsible for a particular behaviour. The brain is a complex network of hundreds of discrete, specialized areas, but none of these acts on its own, and all behaviours are the result of the cooperative activity of many areas. Thus, brain area A, which was activated in our hypothetical fMRI study, is probably involved in other mental processes and is also likely be activated during other tasks.

These criticisms have led some neuroscientists to refer disparagingly to fMRI as 'blobology', and to liken the method to phrenology, the 19th-century pseudoscience that aimed to relate mental functions with the shape of the skull.

The condensed idea
We can measure the structure and function of the living brain

47 Decoding

Researchers can now 'read' brain activity and decode it to predict people's conscious experiences and mental states. This helps them to understand how the brain processes information and could contribute to the production of devices to aid people suffering from paralysis, but it also raises ethical concerns about mental privacy.

I n the past 15 years, neurotechnologies such as functional magnetic resonance imaging (fMRI) have advanced to the point where they can now be used to predict certain mental states, such as what a person is seeing or hearing, from their brain activity. These technologies, which are becoming more sophisticated by the day, help researchers to gain a better understanding of how the brain processes different types of information, and will eventually enable them to develop brain–computer interfaces that can help paralysed patients move and communicate. In principle, they could eventually make it possible to decode a person's thoughts as well, and this raises concerns about invasions of mental privacy.

LOOKING INTO THE MIND'S EYE
Early work carried out in the 1990s showed that brain activity can predict which category of object a person is looking at. This was based on new knowledge, obtained from fMRI studies at the time, that the brain contains discrete, specialized areas that respond specifically to particular types of visual stimuli. The fusiform face area (FFA), for example, fires strongly in response to faces (*see page 38*), but weakly in response to other categories of

TIMELINE

1959	1990s
David Hubel and Torsten Wiesel's experiments on orientation selectivity in the visual cortex of cats	Early fMRI studies predict which object a person is looking at from their brain activity

objects, such as buildings or animals, whereas the parahippocampal place area (PPA) reacts most strongly to images containing houses or visual scenery.

When given information about the levels of activity in these areas, researchers can therefore accurately predict which of the two different types of stimuli a person is looking at, at any given time. And because the FFA and PPA are several centimetres apart in the brain, the functional brain images reliably reveal which of the two is more active while people actually view the images, and this information can also be used to predict accurately the category of object a person is looking at.

MOVIES IN THE MIND

Japanese researchers made a significant advance in decoding about five years ago. They showed people a series of pictures while using fMRI to record the activity in their primary visual cortex (*see box*). Until now, decoding visual

Angling for answers

In the late 1950s, David Hubel and Torsten Wiesel performed a series of experiments that revealed the properties of neurons in the primary visual cortex. They inserted electrodes into the visual cortex of a cat, so that they could record the responses of single neurons to patterns projected onto a screen. They found groups of cells that are very precisely tuned to lines at specific angles, and others that are tuned to lines of a specific angle moving in a specific direction. They also found that cells tuned to the same orientation are arranged in vertical columns, and that the columns are arranged in a systematic, orderly fashion, so that cells in each successive column across the surface of the brain are tuned to increasingly larger angles. Researchers can now decode the activity of these cells to predict and reconstruct still and moving images being seen by a person.

2008	2011	2012
Yukiyasu Kamitani and colleagues reconstruct visual images from primary visual cortical activity	Jack Gallant and team reconstruct moving images from primary visual cortical activity	Bob Knight and colleagues translate brain activity into words

experiences had been limited to images that had been seen before. The studies involved comparing visual cortical activity associated with each one, showing people the same images again, then predicting which one they were looking at from the brain-activity pattern they observed. In this case, though, the researchers recorded primary visual cortical activity associated with one set of pictures, then showed the participants a completely new set of images. They could then decode their brain activity and 'reconstruct' the images they were looking at.

WE ARE OPENING A WINDOW INTO THE MOVIES IN OUR MINDS.

Jack Gallant, 2011

Several years later, researchers in California took this one step further. They scanned study participants' brains while showing them a series of YouTube clips, focusing not only on the primary visual cortex, but also on secondary and tertiary visual cortical areas. They then scanned the participants again as they watched a completely different set of clips, and decoded the visual cortical activity to reconstruct what they were seeing. The reconstructed moving images had a low resolution and were of poor quality, but were easily recognizable.

LISTENING IN

While most of this work looks at brain activity associated with visual perception, researchers have also made progress in decoding the activity associated with comprehending and producing speech. The most advanced research in this area involves implanting electrodes into the brains of epilepsy patients who are being evaluated before undergoing neurosurgery.

In early 2012, researchers in California used this technique to decode the brain activity associated with the processing of heard words. They played prerecorded words to 15 patients, while recording activity in the superior temporal gyrus, which is involved in the intermediate stages of speech processing. They then used a computational model to extract key features of spoken words, such as the time period and volume changes between each syllable, from the brain activity. They could then 'translate' this information back into sounds, to give crude playbacks of the words the patients had heard.

Several months later, another research group in California used the same technique to record and decode the activity associated with articulating vowels from single cells and groups of neurons in the frontal and temporal lobes. The ability to decode this activity not only helps researchers understand how the brain produces speech, but will also be beneficial for the development of brain–computer interfaces (BCIs – *see page 192*) that help paralysed people communicate.

FUTURE FEARS

Brain scanning can be used to decode many other types of mental states. For example, researchers have used it to distinguish between true and false memories, and to predict which of two actions a person is going to perform before they perform it. In the future, it could potentially be used to predict or reveal sensitive personal information, such as personality traits, consumer product preferences or the likelihood that someone will succumb to neurological disease or drug abuse.

Who should, or could, have access to such information? And what if, at some point in the future, people are made to undergo brain scanning in order to reveal this type of personal information against their will? These are major ethical concerns that have to be weighed up against the potential benefits of the technologies.

The condensed idea
Brain activity predicts mental states

48 Brain–computer interfacing

Recent advances have led to the development of devices that can read the brain's electrical activity and translate it into signals which control external machinery. These brain–computer interfaces are commercially available in the gaming industry, and could enable people with spinal cord injuries to control prosthetic limbs or regain control of paralysed limbs.

Just 15 years ago, the idea of using the power of thought to control external devices sounded like science fiction, but brain–computer interfaces (BCIs) now make this possible. BCIs are devices that can decode the neuronal activity and translate it into command signals to control a machine such as a robotic arm. Essentially, they involve wiring up the brain to a computer, either invasively via implanted electrodes, or non-invasively using an electroencephalography (EEG) headset, so that the brain activity associated with the planning of voluntary movements can be rerouted to a machine.

BCIs are based on recent advances in neuroscience, computer science and microelectronics. In the past 30 years, neuroscientists have made a great deal of progress in understanding how groups of neurons in the motor cortex produce movement, largely from studies in which electrodes were implanted directly into the brains of monkeys. At the same time, researchers have developed microwire electrodes that can record brain activity with greater precision, and computer algorithms that decode and translate the activity have become increasingly sophisticated.

TIMELINE

1929	1950s	2005
Hans Berger develops electro-encephalography (EEG) for recording the electrical activity of the brain	Development of the first cochlear implants	John Donoghue and colleagues demonstrate the use of the BrainGate BCI by a quadriplegic patient

In addition, advances in robotics have led to the development of sophisticated artificial limbs with independently moveable digits, raising the possibility that these could eventually be used in conjunction with BCIs to restore movement to severely paralysed patients. Meanwhile, several electronics companies have released cheap BCIs that can be used to control computer games.

THE POWER OF THOUGHT

One brain–computer interface currently being tested in human beings is the BrainGate Neural Interface System. The device, developed by researchers at Brown University, consists of an array of 96 silicon microelectrodes, each just 1mm in length and thinner than a human hair, attached to a cable that connects it to a computer. In 2005, a 25-year-old quadriplegic man named Matthew Nagel became the first recipient of the device. Five years earlier, Nagel had suffered a vicious knife attack that severed his spinal cord and left his arms and legs paralysed. The researchers implanted the electrode array into his motor cortex, and trained him to use it. Within minutes, Nagel had learned how to control the movements of a cursor on the computer screen by merely thinking through the motions, enabling him to send emails and control a television set. The device was then attached to a robotic arm, which Nagel could control to perform rudimentary movements.

THIS IS SCIENCE FICTION COMING TO LIFE.
Brazilian scientist Miguel Nicolelis, 2008

The BrainGate 2 clinical trial is now underway. It currently includes seven stroke patients who are completely paralysed and unable to speak, but the researchers hope to increase this number to 15. In 2012, they reported that two of the patients enrolled in the trials can use the device to command a robotic arm to perform complex three-dimensional movements. One patient, who has been paralysed for over 15 years and has had the electrodes implanted for five years, can command the arm to grasp a water bottle, and pull it towards her mouth to drink from it through a straw.

2007
NeuroSky releases NeuroBoy, a computer game controlled by an EEG-based headset

2010
Guger Technologies releases an EEG-based BCI that enables paralysed patients to type

2011
NeuroSky launches $100 MindWave EEG headset

Despite these remarkable achievements, the BCIs available at the moment are still very crude, and there are a number of obstacles, which researchers hope to overcome in the near future. First, they are very cumbersome because the electrode arrays are attached to a computer by thick cables, and implanting them carries the risk of infection. Furthermore, the current devices are still severely limited in their ability to translate a continuous stream of complex neural activity, and they also have to be monitored constantly by technicians.

In the future, electrode arrays will be made from biocompatible materials that can remain in place for even longer periods of time, and algorithms that translate greater amounts of neural activity from larger numbers of neurons will be developed. BCIs will transmit multiple electrical signals wirelessly, and

Cochlear implants

Cochlear implants are electronic devices that can be surgically implanted into the inner ear to improve hearing in deaf and partially deaf people. They consist of an external and an internal component. The external component sits behind the ear, and consists of a microphone that picks up sound from the environment, a processor that selects speech from the sounds entering the microphone, and a transmitter and receiver/stimulator, which receives sounds from the microphone and converts them to electrical impulses. The internal component is an electrode array implanted into the cochlea; this receives the impulses from the transmitter and relays them to the auditory nerve.

Cochlear implants do not restore normal hearing, but enable people who use them to hear and understand speech sounds. They differ from hearing aids, which merely amplify environmental sounds. Cochlear implants were among the earliest neuroprosthetic devices; since they were developed in the 1950s, approximately 220,000 people have received them worldwide.

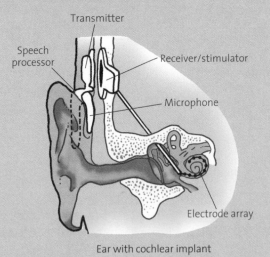

Ear with cochlear implant

prosthetic limbs will be fitted with sensors that provide sensory feedback to the user. This will make artificial limbs feel more like a part of the body, and will enable users to control them more accurately.

COMMERCIAL BCIS

Several electronics manufacturers now offer cheap, non-invasive BCIs that use EEG headsets to record brain waves. The first commercial product was launched by NeuroSky, a California-based company, in 2007. The device was sold with a game called *The Adventures of NeuroBoy*, certain elements of which could be controlled by the headset. Subsequently, NeuroSky partnered with the toy manufacturer Mattel to produce the best-selling consumer BCI product to date – a game called *Mindflex*, which they claim uses brain waves to steer a ball through an obstacle course.

Some of the applications of commercial BCI products have real practical purposes. Recently, for example, an Austrian company called Guger Technologies released a system that allows paralysed people to type, by selecting letters one at a time on a grid. Others are more whimsical. A Japanese company called Neurowear has launched a range of brain-wave-controlled fashion accessories, including a pair of fluffy cat ears that point upwards when the user is concentrating and downwards when they are relaxed. Artists have used EEG-based BCIs to create interactive installations that convert people's brain waves into moving visual displays, and musicians have used them to transform brain-wave patterns into sounds.

Such commercial products will become increasingly available in the years to come as the cost of the technology continues to decrease. The advances in these technologies and their reduced cost will undoubtedly be of huge benefit to those developing BCIs for therapeutic use, too.

The condensed idea
Machines can translate thoughts into actions

49 Neuroscience and the law

Neuroscience is beginning to have profound implications for the legal system. Brain-imaging data are increasingly being admitted as evidence in courts of law, to bolster claims of diminished responsibility and to distinguish truth from lies, while memory research shows that eyewitness testimonies can be highly unreliable. These advances pose major challenges to how we convict and punish criminals.

The law is concerned with distinguishing between innocent and guilty suspects, and punishing those found guilty of committing crimes by regulating their behaviour. The brain controls all of our behaviours, so it follows that advances in neuroscience will be relevant to the law.

Indeed, brain research is already beginning to inform and influence the legal process, in three main ways. First, it raises questions about free will and criminal responsibility; secondly, it raises the possibility of using neurotechnologies to discriminate between the guilty and the innocent; and thirdly it calls into question the validity of eyewitness testimonies.

ARE WE ALWAYS RESPONSIBLE?
The legal system places great emphasis on the concept of responsibility. People are held accountable for their actions, insomuch as they are free to act. Neuroscience is beginning to change our notion of free will, and a growing

TIMELINE

1932	1974	1992
Publication of *Remembering*, Frederic Bartlett's classic book on the nature of memory	Elizabeth Loftus publishes her study of leading questions	Herbert Weinstein's murder charge reduced to manslaughter on the basis of brain-imaging data

number of cases involve evidence of brain abnormalities to argue that a defendant's culpability should be mitigated.

The precedent was set in the early 1990s, in the case of 65-year-old advertising executive Herbert Weinstein, who was charged with strangling his wife. Weinstein's lawyer argued that he should not be held responsible for his actions because of a cyst in his brain that impaired his mental faculties. In light of this, Weinstein's charge was reduced from murder to manslaughter.

Another example is the 2000 case of an American man who suddenly acquired paedophiliac behaviours. The man, a schoolteacher then in his 40s, began visiting prostitutes and collecting child pornography, and eventually started making subtle sexual advances towards his 12-year-old stepdaughter. His wife got wind of this, and he was subsequently arrested and charged with child molestation.

While awaiting sentencing, the man complained of worsening headaches, and these became so unbearable that he was admitted to an emergency room the night before he was due to be sentenced. An MRI scan revealed that he had an egg-sized tumour in his right orbitofrontal cortex, a part of the brain involved in decision-making and social behaviour.

Leading questions

In the 1970s, psychologist Elizabeth Loftus performed a series of simple experiments that convincingly demonstrated how leading questions can influence our memories of events. She showed participants video footage of a car accident, split them into groups, then asked each group a question about the incident in the footage. One group was asked, 'How fast were the cars going when they *bumped* into each other?' while the other was asked, 'How fast were the cars going when they *smashed* into each other?' These subtle differences in phrasing influenced the participants' answers – those asked the first question consistently gave lower estimates of the cars' speed than those asked the second. The findings have obvious implications for how suspects are questioned during police investigations, and how eyewitnesses are cross-examined in the court room.

2007

The Law and Neuroscience Project is established with $10m funding from the MacArthur Foundation

2008

An Indian judge convicts someone of murder on the basis of brain-scanning data

Neurosurgeons removed the tumour, and the man's inappropriate sexual behaviours immediately disappeared. About a year after the surgery, however, his paedophiliac desires returned, and he started secretly viewing child pornography again. Another brain scan showed that the tumour had grown back – evidently, a portion of it had been missed during the first operation. The surgeons operated once again to remove it, and the man's behaviour returned to normal.

DETECTING LIES AND GUILTY KNOWLEDGE

Research interest in the brain mechanisms of deception has increased in the past decade, and at least two American companies now offer fMRI-based lie-detection services to the legal profession. Most neuroscientists agree, however, that we still do not know enough about the brain to distinguish lies from truth on the basis of brain activity, and the general consensus is that lie detection using fMRI scans is no more reliable than the traditional lie-detector test, or polygraph.

Related to this is the so-called 'guilty-knowledge' test, which some researchers claim can be used to determine whether a suspect is concealing knowledge of a crime. In the test, a suspect is shown details of the crime scene while EEG is used to measure the electrical activity of their brain. The electrodes can pick up a specific brain-wave pattern called the P300, which occurs in response to meaningful stimuli.

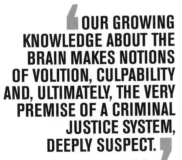

OUR GROWING KNOWLEDGE ABOUT THE BRAIN MAKES NOTIONS OF VOLITION, CULPABILITY AND, ULTIMATELY, THE VERY PREMISE OF A CRIMINAL JUSTICE SYSTEM, DEEPLY SUSPECT.

American neuroscientist
Robert Sapolsky, 2011

The guilty-knowledge test can successfully distinguish between 'guilty' and 'innocent' parties in experimental mock crime scenarios, but it is far less reliable in reality. The real world is far more complex than the carefully controlled laboratory conditions under which the guilty-knowledge test has been evaluated, and the items used in the test could be meaningful to a suspect in some other way, so could elicit a P300 response even if they were not involved in the crime. Furthermore, guilty suspects can employ various countermeasures to control their responses to familiar stimuli. Nevertheless, India set a precedent in 2008, when it became the first country in the world to convict someone of murder based on the test.

DO YOU SWEAR TO TELL THE TRUTH?

The criminal justice system is largely based on eyewitness testimonies, which are routinely used to identify the perpetrator of a crime and to provide information about the criminal event. It's well known, however, that our memories are not as reliable as we like to think they are, and this has profound implications for the use of eyewitness testimonies in courts of law.

In the 1920s, the psychologist Frederic Bartlett performed a series of classic studies that demonstrated how unreliable our memories can be. Bartlett composed a series of short stories, asked his participants to read them, then later asked them to recall as much as possible of the story they had read. He found that people found it extremely difficult to recall the stories exactly, even if they had read them several times.

Bartlett concluded that memory is *reconstructive*, not *reproductive*, in nature. Our memories of events are tainted by our own biases and expectations, which cause errors to creep in to our recollections. Thus, we do not remember events exactly as they occurred; rather, our memories are subtly distorted to fit in to our existing knowledge. Other researchers have since confirmed Bartlett's findings. His work has also been extended to show that memories can easily be manipulated, both inadvertently and deliberately, and that people can be made to believe that completely false memories are actually real (*see page 97*).

The condensed idea
Brain research could change the legal system – for better or worse

50 Neuroethics

Modern neuroscience is changing the way we view ourselves, and advances in our knowledge and the technologies available to us are opening up new possibilities for manipulating the brain and controlling our behaviour. This raises a plethora of ethical issues – neuroscientists are obliged to address these issues, and to help the general public understand their work.

Brain research is progressing at an unprecedented rate, and as our understanding of how the brain operates improves, so too does the ability to manipulate it. Our brains make us who we are, and although our understanding of how the brain works is, at best, still rudimentary, advances in neuroscience are already beginning to redefine the way we view the 'self' and the individual's place within society. Neuroethics is a relatively new interdisciplinary field that seeks to address the wide variety of issues raised by modern brain research.

Some of the issues that neuroethicists deal with – the use of psychosurgery to treat mentally ill people, for example, or using biological samples obtained from experimental subjects without consent – are not new; others – such as the use of stimulants to enhance cognitive function, or employing neuroimaging to decode mental states – have emerged in the past few years; and yet others – the possibility of developing drugs that can erase memories, or the ability to predict crime before it happens – will almost certainly remain unrealized until the distant future, but are already the subject of much debate. Some of the newer neuroethical issues are discussed below.

TIMELINE

1949	1993	2003
António Egas Moniz wins the Nobel Prize for his invention of the lobotomy	UNESCO founds the International Bioethics Committee	The Society for Neuroscience initiates an annual lecture on neuroethics

MENTAL PRIVACY

Your thoughts are your own, inhabiting a private mental world to which none but yourself is privy. But will this always be the case? Neurotechnologies such as fMRI have now reached a point where it is possible to peer into the brain and 'decode' its activity to determine what someone is seeing or feeling, raising concerns about privacy and confidentiality. National security is one area of particular concern: government agencies are increasingly funding this type of research, in the hope of finding behavioural patterns and brain activity 'signatures' that can identify terrorists.

Will your thoughts remain exclusively accessible to you, or will neuroscience eventually enable researchers to read your mind? Although brain-scanning technologies are becoming more sophisticated by the day, many of these claims are exaggerated and unfounded. Neuroscientists can indeed decode your brain activity to determine simple perceptions, but this is unlikely ever to reach the point where they can actually decipher precisely what you're thinking, so your thoughts will in all likelihood remain private for a long time to come.

Psychosurgery: then and now

The lobotomy is a neurosurgical procedure that involves severing the connections between the frontal cortex and underlying structures. It was pioneered in the 1930s by the Portuguese neurologist António Egas Moniz, and developed and then introduced to America by Walter Freeman in the 1940s. Lobotomy is a very crude procedure that produced mixed results. It was, however, hailed as a miracle cure for mental illness. In its heyday during the 1940s and 50s, the procedure was performed on tens of thousands of people in the USA and Europe, before being superseded by anti-psychotic drugs. Lobotomies are no longer performed, but several years ago it emerged that doctors in China are using brain surgery to treat mentally ill patients. They are also treating drug addicts and alcoholics by surgically removing the nucleus accumbens, or 'pleasure centre'. This is considered highly unethical, because neurosurgery is irreversible, and whether the patients give consent is not clear.

2006
Founding of the International Neuroethics Society

2007
The Wall Street Journal exposes an 'epidemic' of psychosurgery in China

2009
Adrian Raine and colleagues find that poor fear conditioning at age three predicts criminality 20 years later

IS THE BRAIN A CRYSTAL BALL?

Neuroscientists can also use fMRI to predict certain simple aspects of behaviour with a good degree of accuracy, and some researchers now claim that they can identify children who will go on to become drug users, criminals or psychopaths on the basis of their brain activity. For example, researchers at the University of Pennsylvania scanned the brains of 41 convicted murderers, all of whom had pled 'not guilty' for reasons of insanity. They found that those who had acted out of impulse, but not those who had planned their crimes, had a reduced level of glucose metabolism in the prefrontal cortex. The same researchers subsequently reported that the amygdala is, on average, 18 per cent smaller in psychopaths than in others, and that poor fear conditioning in three-year-old children is associated with criminality 20 years later.

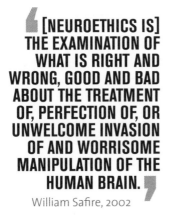

[NEUROETHICS IS] THE EXAMINATION OF WHAT IS RIGHT AND WRONG, GOOD AND BAD ABOUT THE TREATMENT OF, PERFECTION OF, OR UNWELCOME INVASION OF AND WORRISOME MANIPULATION OF THE HUMAN BRAIN.

William Safire, 2002

Will we eventually reach a point at which interventions are forced upon people to prevent undesirable behaviours that have not yet manifested themselves? How would you feel if, at some point in the future, you were told that your five-year-old son is likely to become an alcoholic or a criminal, and that he should be treated immediately to minimize the likelihood of that happening? And if it ever becomes possible to make such predictions accurately, who should have access to the information? Predictions about future behaviour patterns are based on statistical analyses, which show that certain patterns of brain activity are associated with particular behaviours within large groups of people, but these are far less accurate at predicting the behaviour of any one person, because of the variations in brain structure and function that exist between individuals. Nevertheless, this is the type of research that can easily be abused by overzealous social policy-makers – so addressing this issue, and particularly the shortcomings of the work, is vital.

BURNING ISSUES

These are just some of the issues that have emerged in recent years. Neuroethics addresses many other questions, including:

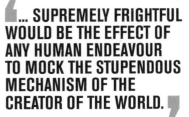

- Should dopamine-based drug therapies be offered to patients with Parkinson's disease if they may produce unwanted side effects such as compulsive gambling?
- Should professors and students use stimulant drugs to enhance their academic performance?
- Should brain stimulation be used to enhance cognitive functioning in healthy people?
- What should researchers do if they find signs of brain damage while scanning someone's brain as part of a study?
- Does brain damage make someone less responsible for their actions and, if so, what are the implications for the criminal justice system?
- Should surgical amputation be offered to people suffering from body integrity identity disorder (*see page 57*)?

... SUPREMELY FRIGHTFUL WOULD BE THE EFFECT OF ANY HUMAN ENDEAVOUR TO MOCK THE STUPENDOUS MECHANISM OF THE CREATOR OF THE WORLD.
Mary Shelley, 1831

READ ALL ABOUT IT

Because of issues such as these, neuroscience is increasingly becoming a part of everyday life, and the general public's interest in the subject has grown accordingly. At the same time, however, ill-informed commentary on the subject is rife. Many neuroethicists therefore now believe that researchers have a professional obligation to reach out to the public, in order to explain the possible implications of their research for society, as well as its limitations.

The condensed idea
Brain research poses
major challenges for society

Glossary

Action potential Electrical signal that travels along nerve fibres and carries information.

Amnesia The inability to remember past experiences or form new memories.

Amygdala Small, almond-shaped structure in the medial temporal lobe, involved in emotions such as fear.

Aphasia Impairment in the ability to produce or understand speech, often due to brain damage caused by a stroke.

Astrocyte Star-shaped glial cell, which provides nutritional support for neurons and participates in information processing.

Autonomic nervous system Branch of the nervous system that controls involuntary functions such as breathing and heart rate.

Axon Single nerve fibre emanating from the cell body of a neuron, which carries impulses to other cells.

Basal ganglia Large set of subcortical structures involved in functions such as voluntary movement and emotions.

Brain stem The midbrain, pons and medulla, which connect the spinal cord to the cerebral cortex.

Central nervous system (CNS) The brain and spinal cord.

Cerebellum The 'little brain', which controls balance and coordinates movements, and also plays a role in thoughts and emotions.

Cerebral cortex Thin, folded structure on the outside of the brain, responsible for higher mental functions.

Cochlea Part of the inner ear containing hair cells that detect movements created by sound waves.

Corpus callosum Massive bundle of nerve fibres connecting the left and right hemispheres of the brain.

Critical period Restricted period of development during which the brain is highly sensitive to experience and sensory stimulation.

Dendrites Branched fibres that carry electrical signals to the cell body of a neuron.

Exocytosis Process by which neurotransmitters are released at synapses.

Fasciculus Bundle of nerve fibres.

Fissure Deep furrow separating different regions of the brain.

Frontal lobe Area at the front of the brain involved in movement and complex mental functions such as planning and decision-making.

Ganglion Cluster of nerve cells that perform the same or similar functions (plural: ganglia).

Gap junction Electrical synapse, which allows the direct flow of electrical signals between neurons.

Glia One of two classes of cells found in the nervous system. There are several different types of glia, each of which performs a specialized function.

Grey matter Nervous tissue containing the cell bodies of neurons.

Growth cone The end of a growing nerve fibre, which detects navigation signals in the environment.

Gyrus Ridge in the cerebral cortex (plural: gyri).

Hippocampus Part of the medial temporal lobe involved in memory and spatial navigation.

Long-term memory Memory that persists for days, months or years.

Long-term potentiation Process by which the connections between neurons are strengthened.

Microglia The brain's immune cells, which clear up cellular debris and devour microbes.

Motor neuron Nerve cell in the brain or spinal cord, involved in planning and executing voluntary movement.

Myelin Fatty tissue that insulates nerve fibres and increases the speed of nervous impulses.

Nerve A bundle of axons in the peripheral nervous system.

Neural stem cell Undifferentiated cell found in the embryo, which can generate all the cell types of the nervous system.

Neuromuscular junction Site at which nerve terminals of spinal motor neurons send signals to muscle cells.

Neuron One of two classes of cells that make up the nervous system. Neurons exist in hundreds or thousands of different types.

Neurotransmission Process by which neurons communicate with each other using chemical messengers.

Neurotransmitter Molecule used by neurons to communicate with each other.

Node of Ranvier Gap in the myelin sheath, at which nervous impulses are generated.

Occipital lobe Region at the back of the brain containing many areas specialized for processing visual information.

Oligodendrocyte Type of glial cell found in the central nervous system, which produces myelin.

Parietal lobe Brain region located behind the frontal lobe, which combines information from the different senses.

Peripheral nervous system Component of the nervous system that lies outside the brain and spinal cord.

Receptor Protein embedded in the nerve cell membrane, to which neurotransmitter molecules bind.

Schwann cell Type of glial cell found in the peripheral nervous system, which produces myelin.

Somatosensory cortex Part of the parietal lobe that receives information from the skin and contains an orderly map of the body.

Somatotopy Orderly representation of the skin surface within the somatosensory cortex.

Spinal cord Bundle of nerve fibres that carries nervous impulses between the brain and the body.

Substantia nigra Part of the midbrain containing neurons that synthesize the neurotransmitter dopamine.

Sulcus Groove between two gyri in the cerebral cortex.

Sylvian fissure Prominent groove separating the frontal and temporal lobes of the brain.

Synapse Junction between two neurons, at which neurotransmission takes place.

Synaptic vesicle Tiny spherical structure found at nerve terminals, which contains neurotransmitter molecules.

Temporal lobe Region at the side of the brain, containing areas specialized for language and memory.

White matter Nervous tissue containing nerve fibre bundles.

Index

Quercus Editions Ltd
55 Baker Street
7th Floor, South Block
London
W1U 8EW

First published in 2013

A catalogue record of this book is available from the British Library

ISBN 978 1 78087 910 9

Printed and bound in China

10 9 8 7 6 5 4 3 2 1

All illustrations by Patrick Nugent except p.34 © Natural History Museum, London/Science Photo Library